Mangey Ram and Suraj B. Singh (Eds.)
Soft Computing

De Gruyter Series on the Applications of Mathematics in Engineering and Information Sciences

Edited by
Mangey Ram

Volume 1

Soft Computing

Techniques in Engineering Sciences

Edited by
Mangey Ram and Suraj B. Singh

DE GRUYTER

Editors
Mangey Ram
Graphic Era Deemed to be University
Department of Mathematics, Computer Science and Engineering
566/6 Bell Road
248002 Clement Town, Dehradun, Uttarakhand, India
drmrswami@yahoo.com

Suraj B. Singh
G. B. Pant University of Agriculture and Technology
263153 Udham Singh Nagar, Pantnagar, Uttarakhand, India
drsurajbsingh@yahoo.com

ISBN 978-3-11-062560-8
e-ISBN (PDF) 978-3-11-062861-6
e-ISBN (EPUB) 978-3-11-062571-4
ISSN 2626-5427
DOI https://doi.org/10.1515/9783110628616

Library of Congress Control Number: 2020936387

Bibliographic information published by the Deutsche Nationalbibliothek
The Deutsche Nationalbibliothek lists this publication in the Deutsche Nationalbibliografie; detailed bibliographic data are available on the Internet at http://dnb.dnb.de.

Cover image: MF3d/E+/Getty Images
Typesetting: Integra Software Services Pvt. Ltd.
Printing and binding: CPI books GmbH, Leck

www.degruyter.com

Preface

Conventional computing or hard computing often depends on a precisely stated analytical model and many times involves a lot of complexity during computation along with high computation time. Nowadays, there are a lot of real-world engineering problems, which cannot be studied precisely due to the presence of impreciseness and uncertainties. To tackle these types of problems, soft computing techniques have found wide applications and have been proved to be a powerful *problem-solving* methodology because of their strong learning and cognitive ability and good tolerance of uncertainty and imprecision. Basically, they are an extension of heuristics and help in solving typical problems that are hard to model mathematically. Moreover, they are easy to accommodate with changed scenario and can be executed with parallel computing. Needless to say, soft computing techniques are an emerging approach, which includes techniques such as fuzzy logic, evolutionary computing, artificial neural network, and applied statistics. An interesting fact to be noted down is that they are actually distinct applied techniques to solve a wide range of problems but when applied in collaboration help a lot in solving complicated problems easily or relatively easily. They have wide applications in fields such as machine intelligence, computer vision, VLSI design, medical diagnosis, pattern recognition, network optimization, and weather forecasting.

Keeping aforementioned facts in view, we have tried to edit a book comprising some of the contemporary researches in different real-world problems where soft computing techniques have been applied. The book consists of ten chapters in which the following studies are conducted:

- Chapter 1 focuses on the application of neural network models for freezing of gait detection and prediction in Parkinson's disease in which authors develop a user-independent freezing of gait detection and prediction model that would go along with nonpharmacological treatments.
- Chapter 2 examines a new fuzzy de novo programming approach for optimal system design, which proposes a new approach for solutions of de novo programming problems in fuzzy environment. This approach is built on the approach by Li and Lee (1993), which uses positive and negative ideal solutions.
- Chapter 3 discusses a probabilistic bilevel programming in Stackelberg game under fuzzy environment in which researchers develop a computational algorithm to solve a stochastic bilevel programming in Stackelberg game using fuzzy optimization technique.
- Chapter 4 studies intuitionistic fuzzy trigonometric distance and similarity measure and their properties. In this chapter, the concept of intuitionistic fuzzy sets is introduced. Similarity and distance measures between intuitionistic fuzzy sets are also explained and extended these measures to intuitionistic fuzzy sets. Some new trigonometric similarity and distance measures of intuitionistic fuzzy

https://doi.org/10.1515/9783110628616-202

sets are studied and it is shown that an intuitionistic fuzzy distance measure satisfies the required identities of intuitionistic fuzzy similarity measures.
- Chapter 5 develops a mathematical model through transmutation of activation energy function. This work incorporates the different perspective of modeling of activation energies through ingression of an additional parameter of distribution function to increase the flexibility and controlling ability of modeling by the linear mixing.
- Chapter 6 addresses forecasting of air quality parameters using soft computing techniques. In this chapter, authors studied the application of soft computing techniques of artificial neural networks and genetic programming for forecasting of air quality parameters a few time steps in advance.
- Chapter 7 examines arithmetic operations on generalized semielliptic intuitionistic fuzzy numbers (IFNs) and its application in multicriteria decision making. In this chapter, authors discuss the generalized semielliptic IFN and performed its arithmetic operations with the help of (α, β) cut method. In this study, they compared the proposed IFN with normal triangular, trapezoidal, semielliptic IFNs.
- Chapter 8 discusses the method for solving intuitionistic fuzzy assignment problem. In this research, authors use a newly proposed centroid concept ranking method for intuitionistic fuzzy numbers and solve assignment problem where assignment costs are taken as triangular intuitionistic fuzzy numbers.
- Chapter 9 focuses on optimization of electric discharge machining process through evolutionary computing and fuzzy multicriteria decision-making techniques. This study intended to explore the best possible set of process parameters, namely, current, voltage, and pulse on time during electric discharge machining process in order to determine the changes in performance characteristics like material removal rate and electrode wear rate for machining of high carbon high chromium die steel workpiece by applying the titanium nitride–coated copper electrode.
- Chapter 10 studies the fuzzy reliability of systems using different types of level $(\lambda, 1)$ interval-valued fuzzy numbers. In this chapter, authors obtain the fuzzy reliability of series, parallel, parallel–series, and series–parallel systems assuming that all components of the system follow different types of level $(\lambda, 1)$ interval-valued fuzzy numbers (both triangular and trapezoidal).

The book will be useful to scientists, engineers, managers, senior and postgraduate students, as well as research scholars.

Acknowledgments

It is well known that besides the editors, many individuals have put much time and energy into the book. First and foremost, we would like to express our gratitude to chapter contributors and reviewers for their timely efforts. We would like to extend extra special thanks and appreciation to the editorial, production, and marketing team at De Gruyter who helped in bringing the project in the final shape.

Mangey Ram,
Graphic Era Deemed to be University,
Dehradun, India

S. B. Singh,
G.B. Pant University of Agriculture & Technology,
Pantnagar, India

Contents

About the editors

Dr. Mangey Ram received his Ph.D. in mathematics with minor in computer science from G. B. Pant University of Agriculture and Technology, Pantnagar, India. He has been a faculty member for around 11 years and has taught several core courses in pure and applied mathematics at undergraduate, postgraduate, and doctorate levels. He is currently professor at Graphic Era (Deemed to be University), Dehradun, India. Before joining Graphic Era, he was a deputy manager (probationary officer) in Syndicate Bank for a short period. He is editor-in-chief of *International Journal of Mathematical, Engineering and Management Sciences* and the guest editor and member of the editorial board for various journals. He is a regular reviewer for international journals, including IEEE, Elsevier, Springer, Emerald, John Wiley, and Taylor & Francis. He has published 200 plus research publications in IEEE, Taylor & Francis, Springer, Elsevier, Emerald, World Scientific, and many other national and international journals of repute and also presented his works at national and international conferences. His fields of research are reliability theory and applied mathematics. Dr. Ram is a senior member of the IEEE, life member of Operational Research Society of India, Society for Reliability Engineering, Quality and Operations Management in India, Indian Society of Industrial and Applied Mathematics, member of International Association of Engineers in Hong Kong, and Emerald Literati Network in the UK. He has been a member of the organizing committee for a number of international and national conferences, seminars, and workshops. He has been conferred with *Young Scientist Award* by the Uttarakhand State Council for Science and Technology, Dehradun, in 2009. He has been awarded the *Best Faculty Award* in 2011; *Research Excellence Award* in 2015; and recently *Outstanding Researcher Award* in 2018 for his significant contribution in academics and research at Graphic Era Deemed to be University, Dehradun, India.

Dr. S. B. Singh is currently professor at the Department of Mathematics, Statistics and Computer Science, G. B. Pant University of Agriculture and Technology, Pantnagar, India. He has around 25 years of teaching and research experience to undergraduate and postgraduate students at various engineering colleges and universities. Prof. Singh is a member of the Indian Mathematical Society, Operations Research Society of India, ISST and National Society for Prevention of Blindness in India, and Indian Science Congress Association. He is a regular reviewer of many books and international/national journals. He has been a member of organizing committee for many international and national conferences and workshops. He is an editor of the *Journal of Reliability and Statistical Studies*. He has authored and coauthored eight books on different courses of applied/engineering mathematics. He has been conferred with five national awards and the best teacher award. He has published his research works at national and international journals of repute. His area of research is reliability theory.

https://doi.org/10.1515/9783110628616-204

List of contributors

Goutam Kumar Bose
Department of Mechanical Engineering
Haldia Institute of Technology
Haldia, West Bengal, India
gkbose@yahoo.com

Alok Dhaundiyala
Szent István University, Mechanical
Engineering Doctoral School
Gödöllő, Hungary
dhaundiyal.alok@phd.uni-szie.hu

Palash Dutta
Department of Mathematics
Dibrugarh University
Dibrugarh, India
palash.dtt@gmail.com

Kanchan Khare
Symbiosis Institute of Technology
Civil Engineering Department
Lavale, Pune, Maharashtra, India
kanchankhare@gmail.com

Pawan Kumar
Department of Mathematics
Statistics & Computer Science
G.B. Pant University of Agriculture & Technology
Pantnagar, Uttarakhand, India
pawankumar44330@gmail.com

S. N. Londhe
Vishwakarma Institute of Information
Technology, Department of Civil Engineering
Kondhwa, Pune, Maharashtra, India
snlondhe@gmail.com

Sumit Kumar Maiti
School of Applied Sciences and Humanities
Haldia Institute of Technology, Purba
Medinipur, West Bengal, India
sumitmaiti123@gmail.com

Amruta Meshram
University of Massachusetts Dartmouth
Dartmouth, USA
ameshram@umassd.edu

Pritam Pain
Department of Mechanical Engineering, Haldia
Institute of Technology, Haldia, West Bengal
India
pritam.me.dscsdec@gmail.com

Bharatendra Rai
University of Massachusetts Dartmouth
Dartmouth, MA, USA
brai@umassd.edu

Sankar Kumar Roy
Department of Applied Mathematics with
Oceanology and Computer Programming
Vidyasagar University, Paschim Medinipur
West Bengal, India
sankroy2006@gmail.com

Laxminarayan Sahoo
Department of Mathematics, Raniganj Girls'
College, Raniganj, West Bengal, India
lxsahoo@gmail.com

Bornali Saikia
Department of Mathematics, Dibrugarh
University, Dibrugarh, Assam, India
bornalisaikia19@gmail.com

D. K. Sharma
Jaypee University of Engineering and
Technology, Raghogarh, Madhya Pradesh
India
dilipsharmajiet@gmail.com

Suraj B. Singh
Department of Mathematics, Statistics and
Computer Science, Govind Ballabh Pant
University of Agriculture and Technology
Pantnagar, Uttarakhand, India
drsurajbsingh@yahoo.com

Shruti Tikhe
DTK Hydronet Solutions, Bavdhan, Pune
Maharashtra, India
shrutitikhe@gmail.com

https://doi.org/10.1515/9783110628616-205

Rajnee Tripathi
Jaypee University of Engineering and Technology, Raghogarh, Madhya Pradesh India
rajneetripathi@hotmail.com

Nurullah Umarusman
Aksaray University, Faculty of Economics and Administrative Sciences, Aksaray, Turkey
nurullah.umarusman@aksaray.edu.tr

Bharatendra Rai, Amruta Meshram

1 Application of neural network to detect freezing of gait in patients with Parkinson's disease

Abstract: Freezing of gait (FOG) consistently reoccurs in the later phases of a patient suffering from Parkinson's disease (PD). Although it is treated with pharmacological treatment, the impact of the medication fades with increasing duration of the disease and thus diminishing the mobility of a patient. This chapter aims at developing a neural network–based classification model that helps to detect FOG episodes in a patient at early stages so that lethal mishaps can be avoided. In this application example, we build user-independent FOG recognition system that would work along in conjunction with nonpharmacological medications. The structured system of developing a neural network–based classification model can be organized into three different stages. The process starts with extraction of suitable features from the dataset. In the subsequent stage, patients are additionally grouped into two clusters depending on the FOG episodes. In the final stage, two neural network models are developed using feedforward network on the two clusters that were formed. The accuracy of the model is computed using sensitivity and specificity.

Keywords: healthcare, Parkinson's disease, freezing of gait, feature extraction, clustering, neural network

1.1 Introduction

Parkinson's disease (PD), caused by Parkinsonism, was first discovered by James Parkinson in 1817. This symptom is observed due to a decrease in the neuromelanin neuron called dopamine causing unusual activity in the cerebrum [1]. Main features of PD are frequently categorized into four groups: trembling, rigidity, akinesia, and posture imbalance [2]. The Hoehn–Yahr score (H&Y) is used to measure different stages of PD in a patient, which are classified into five distinct stages indicating relative disability levels as follows [3, 4]:

Stage 1: Affects only one side of the body with minimal or no functional impairment.
Stage 2: Affects both sides of the body without the loss of balance.

Bharatendra Rai, Amruta Meshram, University of Massachusetts, Dartmouth, USA

https://doi.org/10.1515/9783110628616-001

Stage 3: Impairment is found; functionality is restricted. In this stage, the patient is physically capable of doing work independently, and the disability is mild or moderate.

Stage 4: A higher level of disability, but the patient can still walk or stand without being assisted with a conspicuous concentration difficulty.

Stage 5: Bounded to a bed or a wheelchair, but needs an assistant.

Freezing of gait (FOG) is the short, inconstant loss or an explicit deterioration in the movement of the feet, despite inclination or willingness to walk [5, 6]. Some common symptoms of a FOG episode involve difficulty in executing efficient stepping even when the patient makes an effort to do so, a feeling of difficulty to move legs, and also shivering of legs [6]. About 70% of FOG is observed in a patient during advanced stages of PD, but 26% of FOG is reported in the beginning stages for those patients who have not undergone a levodopa therapy yet [5]. FOG experienced by a patient is mostly a brief episode that can sometimes last more than 1 min [7].

As the mechanism of the freezing experienced by a patient is multifactorial and not well understood, the treatment to find a solution is very challenging [8]. Levodopa is the most efficient drug used for PD [9]. Some other drugs that are available are dopamine agonists, catechol-*o*-methyl-transferase inhibitors, and nondopaminergic agents, but the effect of these drugs does not stay for a longer duration in patients [9].

Although the complete cure of PD is not available at this time, if FOG events can be predicted even a few seconds before they occur for a patient, nonpharmaceutical treatments can be very helpful to avoid fatal accidents. These nonpharmaceutical treatments may take the form of visual, auditory, and tactile cueing provided to the patient [10].

The objective of this chapter is to build a user-independent classification model by application of neural network techniques to enhance FOG identification in PD. To accomplish this objective, the data related to patients are partitioned into two clusters depending on the features and then we develop a feedforward neural network model for each of the two groups.

1.2 Data and approach

In this chapter, we will use Daphnet FOG dataset which is freely available [4]. This dataset is based on experiments that were performed at the Laboratory for Gait and Neurodynamics at the Department of Neurology of the Tel Aviv Sourasky Medical Centers (TASMC). The dataset consists of FOG events captured with the help of wearable accelerometer sensors.

Figure 1.1 provides a representation of the sensor data that will be used in this chapter. In this study, three sensors are utilized to quantify acceleration in three

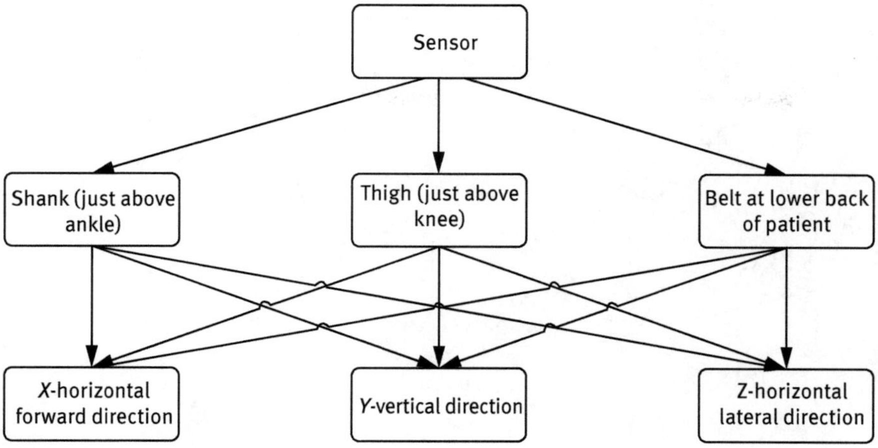

Figure 1.1: Sensor usage flowchart.

dimensions: first appended at the shank, second at the thigh, and third at the lower back. Every sensor collects reading from movements related to three axes: *X*-axis, *Y*-axis, and *Z*-axis. The horizontal forward direction reading is collected using *X*-axis, the vertical direction reading is collected using *Y*-axis, and the horizontal lateral information is collected using *Z*-axis. The dependent variable describes the presence or absence of the FOG. In total, the database consists of ten variables: one dependent and nine independent variables.

Figure 1.2 provides a flowchart depicting the data collection phase. From the flowchart we can observe that in all there are ten patients: three female patients and seven male patients. These ten patients have the disease span varying from 2 to 30 years. The age for ten patients ranges from 59 to 75 years. The patients are further classified as belonging to the "ON" phase of medication and "OFF" phase of medication. At the time of collecting data, each patient completes diverse walking patterns that involve an arbitrary walk including 360° turn, walking in straight line including 180° turn and walking while performing simulation tasks. All these events were recorded using a digital video camera. A physiotherapist is used for noting events during the experiments to determine start time, duration, and end time for the FOG event. In addition, the current activity of the patients is also recorded.

The data collection period involved recording for 8 h and 20 min with an interval of 15–20 ms. Out of the ten patients, eight patients experienced FOG episodes. Overall, 237 FOG episodes were detected during the experiment. Patient 5 encountered as many as 66 FOG episodes, whereas patient 6 experienced 10 FOG episodes which were least among all the patients with FOG. Figure 1.3 represents a boxplot of the duration of FOG events for each patient. Data of patient 4 and patient 10 are excluded from the plot as they did not experience any FOG event. Although patient 6 (disease duration: 22 years) experienced only 10 FOG events, we can observe from

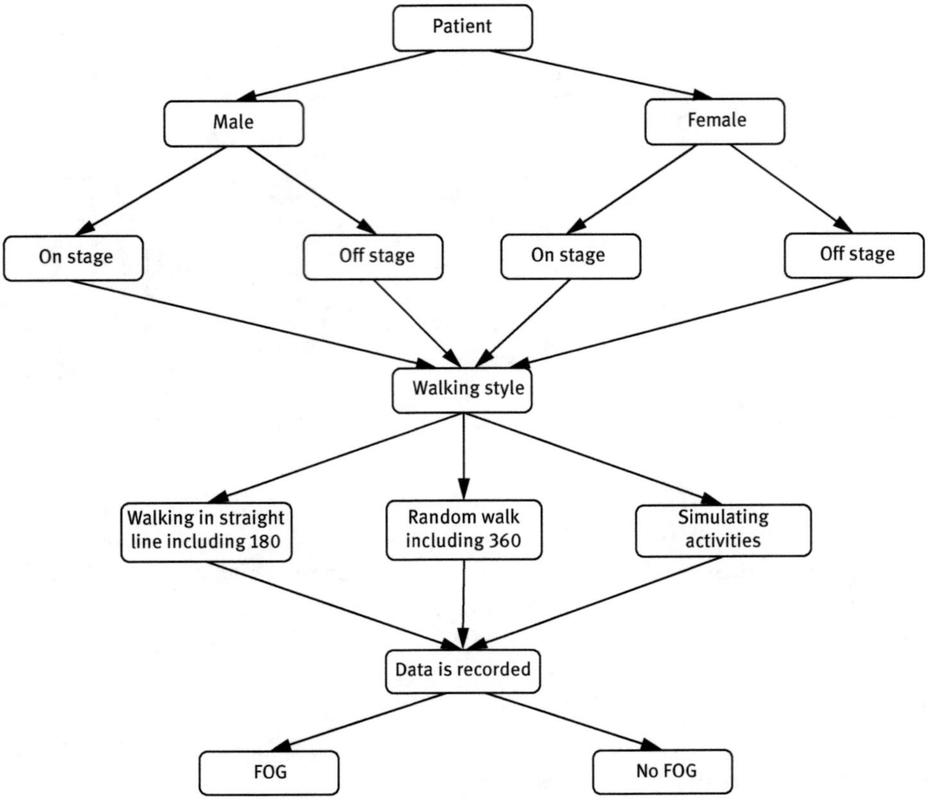

Figure 1.2: The patient data collection process.

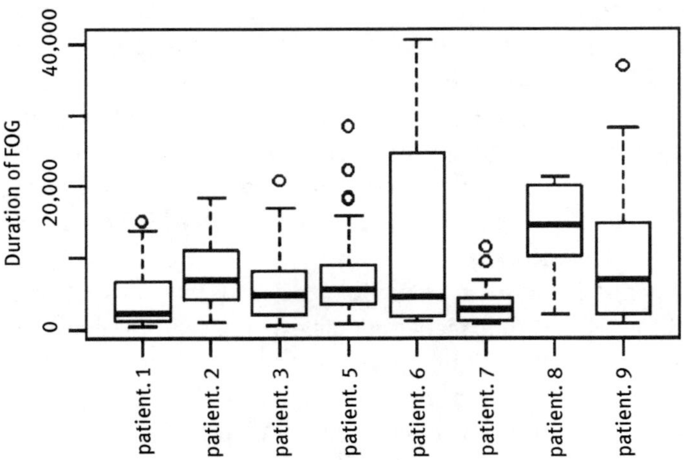

Figure 1.3: Boxplot for duration of FOG events in each patient (Patients 4 and 10 are excluded as they did not experience FOG events).

the boxplot that a period of FOG events was the highest variability (sd = 14.32 s) for this patient. Patient 8 (disease duration: 18 years) also indicates relatively high variability in length of FOG events.

On the other hand, patient-7 (disease duration two years) experienced the least variability (sd = 2.73 s) among all patients. Patient-8 was also observed to have the highest average FOG event duration (mean = 14.34 s), and patient 7 had the least average (mean = 3.38 s). It is to be noted that patient 8 was on the "ON" stage and was the only patient with the highest H&Y score of 4.

1.3 Methodology

This section provides a detailed overview of the three stages of developing a classification model using neural network.

Stage 1: (Extracting essential feature for model building): In this analysis, use of a 50% overlapping window is utilized to derive 11 statistical features. The statistical features are derived from the nine variables and consist of mean: mean1 (lower 10% of the values were trimmed), mean2 (lower 20% of the values were trimmed), minimum, maximum, median, standard deviation, kurtosis, variance, skewness, and mode. Thus, in all we develop 99 features from the data.

Stage 2: (Clustering analysis for model building): In this phase, data with FOG episodes are considered for the analysis and the patients are grouped into clusters depending on their FOG episodes. FOG events using all 99 features were utilized to perform *k*-means cluster analysis. This analysis resulted in the formation of two distinct clusters: patients 1, 2, 3, and 5 are classified as cluster 1, while patients 6, 7, 8, and 9 are grouped as cluster 2.

Stage 3: (Model building): In this stage, we build a neural network model to classify and predict FOG events. In this dataset, the overall percentage of FOG events is 9.7%, which means the percentage of no-FOG event is 90.28%. If we consider the worst scenario where without developing any classification model we make a prediction that all events belong to no-FOG category, then we will be right 90.28% of the time. We can use this number as a benchmark and develop a classification model that has a better performance. In this study, we will make use of sensitivity and specificity to assess the performance of the classification model. Sensitivity is defined as the ratio of FOG episodes that are correctly classified out of all the FOG episodes observed in the data. Whereas specificity is defined as the ratio of no-FOG episodes that are correctly classified out of all the no-FOG episodes observed in the data.

To perform classification of the two clusters, feedforward networks (nntraintool) in MATLAB are utilized to classify inputs according to the target classes. While developing a classification model on the two clusters formed, 70% of the data are used as training data, 15% of the data are used as validation data, and the remaining 15% of the data are used as test data. The representation of cross-entropy versus performance using training, validation, and test data for cluster 1 is depicted in Figure 1.4.

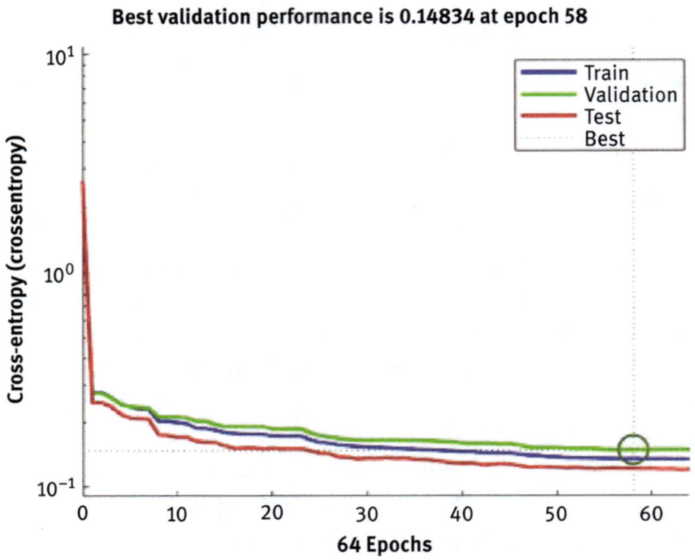

Figure 1.4: Cross-entropy versus performance using training, validation, and test data for cluster 1.

For developing a good classification model we minimize cross-entropy. As shown in the plot, the cross-entropy reduces with the increasing epochs for training, validation, and test data. The dotted green line indicates the best epoch, which in this case is 58 epochs. The performance of the network is measured using four lines that capture train, validation, test, and best situations. If any of the three lines (training, validation, and testing) meet or pass near the best (dotted) line, it symbolizes convergence.

To assess the classification performance of the mode, we develop a confusion matrix. Figure 1.5 provides the confusion matrix for cluster 1.

From the confusion matrix, we can observe that there is consistency in results for training, validation, and test data. For training data, the model correctly classified 8,855 events as no-FOG and 457 events as FOG. The sensitivity and specificity values for the training data are 66.4% and 96.1%, respectively. The sensitivity and specificity values for the validation data are 63.1% and 96.3%, respectively. Similarly, sensitivity and specificity values for test data are 64.4% and 97%, respectively. This consistency

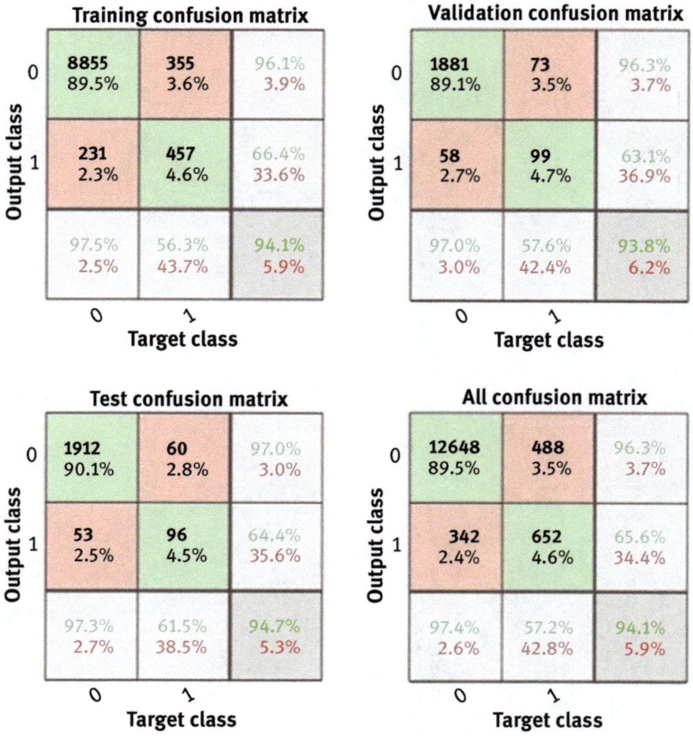

Figure 1.5: The confusion matrix for cluster 1.

in results across training, validation, and test data indicates the absence of overfitting problem. Another observation that we can make from this confusion matrix is about the events that are misclassified. It is observed that the events that belong to FOG category are misclassified as no-FOG at a slightly higher rate compared to a situation where events belonging to no-FOG category are misclassified as FOG. This pattern occurs consistently for the training, validation, and also the test data.

Figure 1.6 represents the receiver operating characteristics (ROC) curve of the feedforward network at epoch 58 for cluster 1.

From the ROC we observe that the classification performance for the training, validation, and test data are again consistently high. The diagonal line in the plot depicts the classification performance with 50% accuracy. And closer the ROC curve to the upper left corner of the plot, better the classification performance.

The representation of cross-entropy versus performance using training, validation, and test data for cluster 2 is depicted in Figure 1.7.

As observed, the cross-entropy reduces with the increasing epochs in training, validation, and test data. The dotted green line indicates the best epoch, which is 75 epochs.

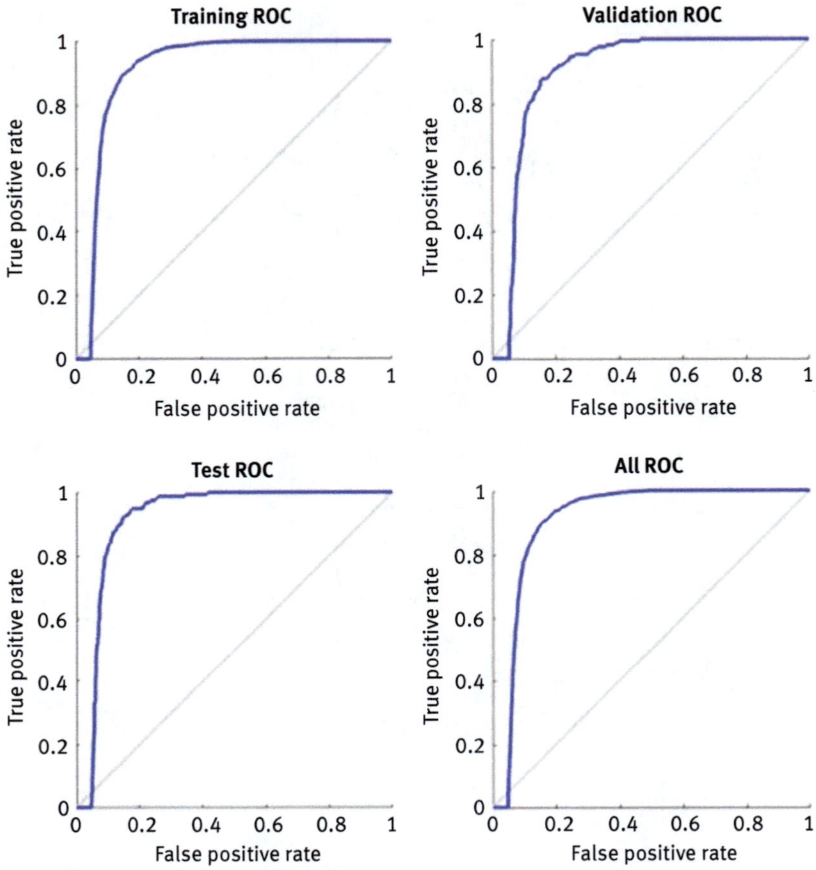

Figure 1.6: The ROC curve for a feedforward network at 58 splits for cluster 1.

Figure 1.8 provides the confusion matrix for classification model with cluster 2.

From the confusion matrix we can observe that the sensitivity and specificity values for the training data are 63.7% and 91%, respectively. The sensitivity and specificity values for the validation data are 71.6% and 91.9%, respectively. Similarly, we observe that the sensitivity and specificity for test data are 74.6% and 90%, respectively. Once again we observe high consistency in classification performance for training, validation, and test data. We also note that the classification performance for cluster 2 is better compared to results obtained with cluster 1. In addition, it is also observed that misclassification of FOG event as non-FOG event is significantly higher compared to situations where non-FOG event is misclassified as FOG event.

Figure 1.9 represents the ROC curve of the feedforward network at epoch 75 for cluster 2.

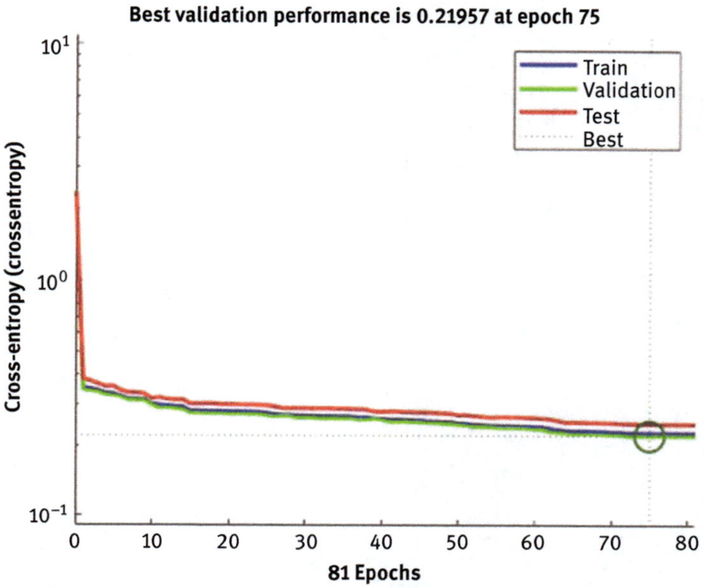

Figure 1.7: Cross-entropy versus performance using training, validation, and test data for cluster 2.

The ROC curves based on training, validation, and test data show consistently high performance compared to the diagonal line representing 50% accuracy.

In this chapter, we classified all patients into two clusters to enable us to develop patient-independent model. The classification model where both training and testing data belonging to the same patient is a user-dependent model, whereas the model where training is performed on a set of patients and testing is performed on another set of patients is a user-independent model [11]. When classification models are made for an individual patient, this model is likely to have better performance. However, such a model is of limited use to another patient. Developing a user-independent model is cheaper and efficient compared to a user-dependent model but accuracy of both models is less. Reasons for such lower accuracy are inconsistency in data due to different walking styles, small size of the dataset, and a limited number of patients.

1.4 Conclusion

A user-independent methodology for detection and prediction of FOG events using combination of clustering and neural network model is illustrated in this chapter. This chapter is divided into three stages: feature extraction, clustering, and model development/evaluation. The dataset contains data from three different sensors

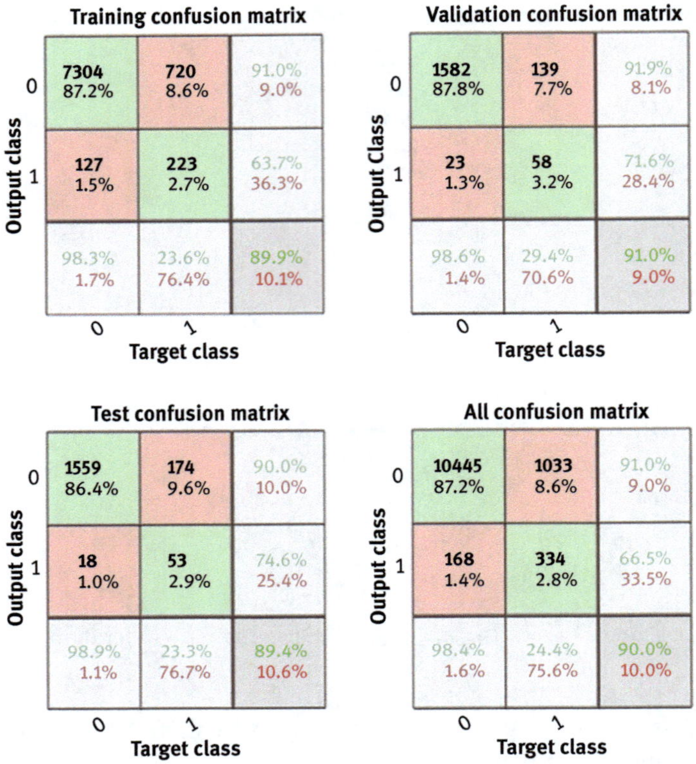

Figure 1.8: The confusion matrix for cluster 2.

that are used to derive 99 variables with the help of 11 statistical numeric summary values. An overlapping window of 50% is utilized to extract the features that were further used to group patients into two clusters using *k*-means clustering. Feedforward neural network classification model is built for each of the two clusters.

In this study, data from all sensors were utilized to perform the experiment. We used neural network–based classification model in this chapter. Further research may involve use of various other machine learning and deep learning methods to improve classification performance. In addition, this dataset used has a significantly lower number of FOG events (about 10%) compared to no-FOG events (about 90%). Such a severe imbalance in the two categories is likely to cause bias in results where no-FOG events are classified with a much higher accuracy compared to accuracy in classifying FOG events. This bias is observed in the classification model that was developed for cluster 1 and cluster 2. In both classification models, accuracy in correctly classifying no-FOG events is higher than 90%. Whereas accuracy in correctly classifying FOG events for classification models for both cluster 1 and cluster 2 has a percentage lower than 75%. This disparity in accuracy for two classes observed due to class imbalance is an interesting problem and needs further investigation.

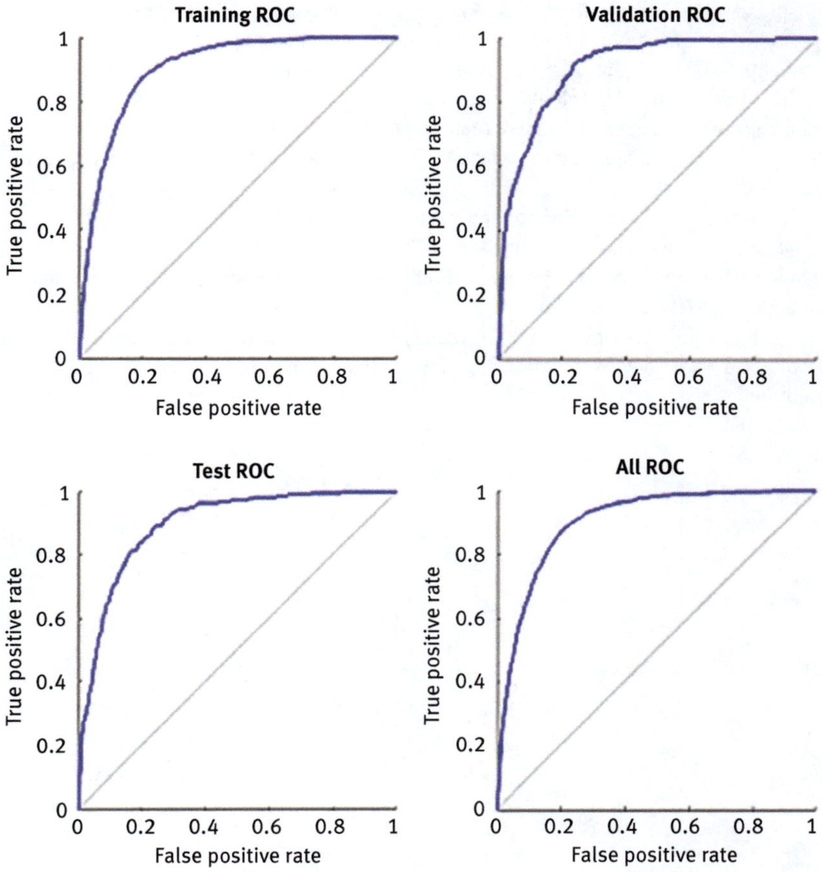

Figure 1.9: The ROC curve for a feedforward network at 75 splits for cluster 2.

References

[1] Fahn S. (2003) Description of Parkinson's disease as a clinical syndrome, Annals of the New York Academy of Sciences, 991, 1–14.

[2] Jankovic J. (2008) Parkinson's disease: clinical features and diagnosis, Journal of the Neurological Neurosurgery Psychiatry, 79(4), 368–376.

[3] Hoehn M.M. & Yahr M.D. (1967) Parkinsonism: onset, progression, and mortality, Neurology, 17(5), 427–427.

[4] Bächlin M., Plotnik M., Roggen D., Maidan I., Hausdorff J.M., Giladi N., & Tröster G. (2010) Wearable assistant for Parkinson's disease patients with the freezing of gait symptom, IEEE Transactions on Information Technology in Biomedicine, 14(2), 436–446.

[5] Heremans E., Nieuwboer A., & Vercruysse S. (2013) Freezing of gait in Parkinson's disease: where are we now?, Current Neurology Neuroscience Reports, 13(6), 350.

[6] Nutt J.G., Bloem B.R., Giladi N., Hallett M., Horak F.B., & Nieuwboer A. (2011) Freezing of gait: Moving forward on a mysterious clinical phenomenon, Lancet Neurology, 10(8), 734–744.
[7] Schaafsma J.D., Balash Y., Gurevich T., Bartels A.L., Hausdorff J.M., & Giladi N. (2003) Characterization of freezing of gait subtypes and the response of each to levodopa in Parkinson's disease, European Journal of Neurology, 10(4), 391–398.
[8] Okuma Y. (2014) Practical approach to freezing of gait in Parkinson's disease, The Practice of Neurology, 14(4), 222–230.
[9] Jankovic J. & Aguilar L.G. (2008) Current approaches to the treatment of Parkinson's disease, Neuropsychiatric Disease and Treatment, 4(4), 743–757.
[10] Boelen M. (2007) The Role of rehabilitative modalities and exercise in Parkinson's disease, Disease-a-Month, 53(May), 259–264.
[11] Delval A., Snijders A.H., Weerdesteyn V., Duysens J.E., Defebvre L., Giladi N., & Bloem B.R. (2010) Objective detection of subtle freezing of gait episodes in Parkinson's disease, Movement Disorders, 25(11), 1684–1693.

Nurullah Umarusman

2 A new fuzzy de novo programming approach for optimal system design

Abstract: De novo programming, which evaluates the concept of optimality from a different point of view, makes it possible to rearrange the number of resources for constraint functions depending on the budget constraint. While the resource amounts of the constraints are used at full capacity, this arrangement improves the performance level of the objective functions. With the development of fuzzy set theory, mathematical programming models can be examined in a fuzzy environment as in other scientific fields. Mathematical programming models are handled in different ways as fuzzy decision variable, fuzzy goal, and fuzzy parameter. The fuzzy goal and fuzzy parameter models are used in the literature for de novo programming, also known as optimal system design. In this study, a new fuzzy de novo programming approach was proposed by using positive and negative ideal solutions. In the proposed approach, using fuzzy unit prices of resources and fuzzy resource amounts of constraints, the fuzzy budget was created. The solution steps of the developed approach are given step by step on an illustrative example accepted in the literature.

Keywords: Multiobjective de novo programming, multiobjective linear programming, fuzzy de novo programming, optimal system design

2.1 Introduction

In classical linear programming (LP) problems, there is a goal to be achieved based on certain decision variables and constraints. The solution of such problems determines the decision variable values as well as the objective function values based on them. On the other hand, it is possible to define the resource amount in each constraint using the values of decision variables. Instead of focusing on objective function values after solving the problem, it may answer significant problems to concentrate on the utilization amount of constraint resources, such as "Are resource amounts utilized optimally?" The answer to this problem would result in the interpretation of all the functions in the problem. When a problem is analyzed in terms of the amount of resource utilization, it mostly leads to an excess amount of resources and/ or a need for more. Either of them is a reason that hinders the optimal utilization of resources for production and also causes a downfall in the performance level of objective function. All constraints must be active for classical LP to determine an

Nurullah Umarusman, Faculty of Economics and Administrative Sciences, Aksaray University, Turkey

https://doi.org/10.1515/9783110628616-002

optimal result. In other words, all constraint resources must be utilized at full capacity. Then, such a problem is named as an optimal model in terms of resource utilization. In multiobjective linear programming (MOLP) problems, the conditions described above are taken to a different dimension with “the increase in the number of goal function.” It is because each goal function is achieved at different values of different variables in the solution of MOLP problems. That would lead to differences in resource utilization amounts of constraints in terms of each objective function.

It is an expected case in MODM problems that constraints have limited amounts and goal functions are not homogeneous. Therefore, there is no concept of “optimization” in MODM (multiple objective decision-making) problems. That is why it is almost impossible for all goal functions to be realized optimally and simultaneously [1]. A significant feature of MODM methods is the concept of “optimality,” which is defined for single objective optimization problems but not valid for MODM problems. Instead of the concept of “optimality,” decision makers therefore use the concept of pareto-optimality (alternatively, efficient, nondominated, or noninferior), proposed by Italian economist Pareto, for MODM problems [2]. The concept of “Pareto-optimality” is named as “pareto-optimal” solutions for MODM problems as defined by Yu [3]. The set number of nondominated solutions is quite high in MOLP problems. In a MODM problem with p-number of goal functions, the concept of “optimality” for any goal function may not be applicable to other p–1 goal functions. The fundamental process in evaluating nondominated solutions is how to determine the proximity to ideal solution points (Cohon, 1978). The methods in MODM classification are used in the solution of MOLP problems. It is possible to reach a satisfactory solution among nondominated solutions by using the interactive methods such as step method, Geoffrion–Dyer–Freinberg method, and Zoints and Wallenius method [4]. Another process in evaluating nondominated solutions is the analysis based on positive and negative ideal solutions. The utility approach, goal programming, interactive approaches, and fuzzy approach models can be applied to reach a compromise solution [5]. Additionally, Tabocanon (1988) described solutions of MOLP problems using global criterion method, interactive methods, and compromise programming methods.

The studies by Zeleny [6, 7] provided a different viewpoint in solving single objective linear programming (SOLP)/MOLP problems. Named as “optimal system design,” this approach enables the reorganization of resource amounts using “budget.” Also known as de novo programming, “optimal system design” can be regarded as a special extension of LP. The most significant feature of de novo programming is that it can reorganize resource amounts of constraints based on the budget constraint. Resource amounts of constraints are utilized at full capacity with this reorganization. It brings an optimal model in terms of resource utilization. Therefore, de novo programming approach may be named as a redesigned SOLP/MOLP model. “De novo” hypothesis may be easily applied to MOLP problems as well. However, “optimal system design” application for MOLP inhabits a long process.

This study proposes a new approach for solutions of de novo programming problems in fuzzy environment. This approach is built on the approach by Li and Lee [8], which uses positive and negative ideal solutions. The proposed approach is analyzed with an illustrative example.

2.2 An overview of the multiobjective de novo programming process

The first study about de novo programming, which designs an optimal system instead of optimizing a given system, was carried out by Zeleny [6]. Today's production and management systems are necessarily becoming more and more flexible. They need to be designed and redesigned rapidly, disassembled and reassembled again, which requires continuous reorganization of resources for ensuring optimality [9]. As all systems are created within their limits, they lack alternatives, options, and design variables in their creation environment. Therefore, when a system is to be redesigned, reconfigured, or optimized, its limitations and boundaries have to be done so as well. It is not sufficient to reconfigure it based on a given system with its own priorities and options. Therefore, system design requires creations of alternatives instead of selection [10]. In contrary to optimizing a given system as standard approaches do, de novo enables constraint variables and works to reach a solution that is more reassuring than those within the system's limitations. This approach must be implemented prior to the onset of production because optimal production is necessarily dependent on optimal utilization of raw materials which is only possible with an optimal production plan (Babic and Pavic, 1996).

2.2.1 De novo programming literature survey

It is possible to classify the scientific articles on de novo programming based on their theoretical and application aspects. This subtitle summarizes the studies that theoretically contribute to de novo programming problems.

As stated above, Zeleny [6] was the first to propose de novo programming approach, which designs an optimal system. Zeleny [7] applied to a classical LP problem. These two studies make up the foundation of de novo programming problem. Zeleny [11] summarized the approach and methodology of de novo programming, which is related to extensive LP and LP-related problems. Implications for solving traditional LP problems are also given in his study. It also provided new knowledge on the "new concept" of optimality and of MCDM. Hessel and Zeleny [12] used the concept of metaoptimum to explain how the system design can be carried out for multiobjective de Novo programming. Zeleny (1989) proposed an ERA (External Reconstruction Approach) method with very simple philosophy. The right side constants of constraints are organized using its specific algorithm. This organization enables the full utilization of constraint constants. It can be regarded as a different aspect of de novo approach. This approach does not use the budget constraint. Zeleny [10] suggested the "optimal path method" to create the design of an optimal system to conduct the solution of a de novo problem through an ideal system design. Shi [13] proposed in his study that several optimum path ratios could be used as alternatives to select the optimal system design to solve multiobjective de novo programming problems. Apart from "optimum path ratios" proposed by Zeleny [10] and Shi [13], there is no certain method to be used in solving multiobjective de novo programming problems. Some studies make use of MODM methods to solve multiobjective de novo programming problems. Tabocanon (1988) solved multiobjective de novo programming problems by using global criterion method, compromise constraint method, and step method. Umarusman [14] demonstrated how multiobjective de novo programming problems could be solved by using minmax goal programming. Umarusman and Türkmen [14] suggested the global criteria method, which relies on positive ideal solutions to solve multicriteria de novo programming problems. Zhuang and Hocine [15] proposed the meta goal programming for the solutions of multicriteria de novo programming problems. Banik and Bhattacharya (2018) suggested weighted goal programming method to solve multiobjective de novo programming problems.

The short literature survey above lists some of the articles that investigate the solution of multiobjective de novo programming problems in classical environments. In addition to classical decision-making process, there are significant studies on de novo programming problems in fuzzy environments as well. Below are the theoretical studies on the de novo programming in fuzzy environments. First, Lai and Hwang [16] made use of Chanas' (1983) nonsymmetric approach to solve a single criterion de novo programming problem. Li and Lee [17] were the first to propose a fuzzy approach to solve a multicriteria de novo programming problem using "the possibility concept of fuzzy set theory." In their study as a continuation of the former, Li and Lee [18] suggested a "two-step fuzzy approach" based on positive ideal solutions and negative ideal solutions. They proposed that their approach is very efficient and can be applied to most multicriteria de novo programming problems. In their next article which is considered to be a continuation of their first study where they analyzed multicriteria de novo programming in a fuzzy logic frame, Li and Lee [8] studied fuzzy goals and fuzzy coefficients simultaneously and proposed a different approach. Bhattacharya and Chakraborty [19] suggested an alternative method for general multiobjective de novo programming problems under fuzzy environment, which uses Luhandjula's compensatory μ_θ-operator. Banik and Bhattacharya (2018) put forward a one-step method to solve multiobjective de novo programming problem, which makes use of min–max goal programming technique where all parameters are fuzzy numbers. Umarusman

[20] suggested for solving multiobjective de novo programming problem in fuzzy setting using Wu and Guu (2001) approach.

2.2.2 Multiobjective de novo programming formulation

Similar to MOLP problems, multiple objective de novo programming problems are handled with more than one and conflicting objective functions based on the same constraint functions. Multiple objective de novo programming is demonstrated mathematically as follows [10]:

$$\text{Max}\, Z_k\colon \sum_{j=1}^{n} C_{kj}x_j$$

$$\text{Min}\, W_s\colon \sum_{j=1}^{n} C_{sj}x_j$$

Subject to (2.1)

$$\sum_{j=1}^{m} A_jX_j \otimes B$$

$$x_j \geq 0,\ k = 1, 2, \ldots, l,\ s = 1, 2, \ldots, r, \text{ and } j = 1, 2, \ldots, n$$

where

$A_j = \sum_{i=1}^{m} p_i a_{ij}$ is the total budget with B utilization, and P_i is the unit price of ith resource. $\otimes$ symbol means that the budget constraint is "≤" or "=" type. This decision rests on the decision maker. "=" allows the total use of budget constraint, whereas "≤" means that it should be less than the current budget.

De novo formulation provides not only the best mixture of outputs but also the best mixture of the inputs (Tabucanon, 1988). One of the advantages of this approach is that it enables the adjustment of resource constraints in multicriteria decision-making problems and that it makes the initial solution more convenient at the same or less cost. Resource amounts of constraints are utilized at full capacity with de novo solution. It enables slack, surplus, and artificial variables to be equal to zero. When the solution of de novo programming problems is investigated from MOLP perspective, positive and negative ideal solutions [21] must be determined for each objective functions. The positive ideal solution set is indicated as $I^* = \{Z_1^*, Z_2^*, \ldots, Z_l^*; W_1^*, W_2^*, \ldots, W_s^*\}$ and negative ideal solution set as $I^- = \{Z_1^-, Z_2^-, \ldots, Z_l^-; W_1^-, W_2^-, \ldots, W_s^-\}$. Then the solution of eq. (2.1) may be done in terms of "optimal path ratio" as proposed by Zeleny [10] and Shi [13], based on positive and negative ideal solutions.

2.3 An approach proposal for fuzzy multiobjective de novo programming

Conventional optimization methods evaluate parameters based on precise data. However, decision makers want to know which changes in parameters would result in which changes in the solution of the problem. It is because the precise parameters used by decision makers are only valid for "that moment." It becomes harder to determine optimal values of goals when short- and long-term changes in parameters are accounted for. The development of fuzzy set theory by Zadeh [22] and further improvements made it possible to acquire more flexible and informative results that present solution by forming given parameters structured as fuzzy ranges. Fuzzy set theory, which helps decision makers in fuzzy decision environments, may be easily applied to multiobjective de novo programming problems. This chapter presents a new methodology based on the proposal by Li and Lee [8], which accounts for both goals and coefficients at the same time.

The model by Li and Lee [8], which presents fuzzy goals and fuzzy coefficients ($\tilde{c}_{kj}, \tilde{c}_{sj}, \tilde{a}_{ij}, \tilde{p}_i$, and $\tilde{B}$) formed by using positive ideal and negative ideal solution information is here solved as a two-step fuzzy approach. Consider the fuzzy multicriteria de novo programming problem:

$$\text{Max}\, \tilde{Z}_k(x) = \sum_{j=1}^{n} \tilde{c}_{kj} x_j$$

$$\text{Min}\, \tilde{W}_s(x) = \sum_{j=1}^{n} \tilde{c}_{sj} x_j$$

Subject to (2.2)

$$\sum_{j=1}^{n} \tilde{a}_{ij} x_j - b_i \le 0$$

$$\sum_{i=1}^{m} \tilde{p}_i b_i = \tilde{B}$$

$$x_j \ge 0,\ i = 1, 2, \ldots, m,\ j = 1, 2, \ldots, n,\ k = 1, 2, \ldots, l, \text{ and } s = 1, 2, \ldots, r.$$

Parameters $\tilde{c}_{kj}, \tilde{c}_{sj}, \tilde{a}_{ij}, \tilde{p}_i$, and $\tilde{B}$ in eq. (2.11) are fuzzy numbers. The fuzzy parameter model proposed by Li and Lee [8] may be summarized as follows: different ranges are used for maximization or minimization goals in eq. (2.11). For a maximization problem, $[f^1, f^0)$ type ranges may be used for the fuzzy coefficients in the goal function. On the other hand, fuzzy coefficients in the goal function must be set in $(f^0, f^1]$ range for the minimization problem. Then, the fuzzy coefficients $\tilde{a}_{ij}, \tilde{p}_i$ are $(f^0, f^1]$ type, and the fuzzy budget $\tilde{B}$ is $(f^0, f^1]$ type. Therefore, an optimal system design at the safety level α is formed according to the highest profit, lower cost, highest amount of available, lowest resources price, and smallest units of materials to be used for each product. Membership functions may be defined for these ranges using linear form. Next, the compromise solution is conducted for goal functions based on the defined α – cut value. According to Li and Lee [8], $(x)^{\alpha}_{\beta}$ be a solution of (2.2) where $\alpha \in [0;1]$ denotes the possibility to which all fuzzy parameters satisfy, and $\beta \in [0;1]$ denotes the compromise to which the solution satisfies the fuzzy goals at given possibility α. λ denotes the overall satisfaction for the solution $(x)^{\alpha}_{\beta}$. When $\alpha = \beta = \lambda$ the optimal solution is obtained. If the results do not satisfy $\alpha = \beta$, choose another α and solve phase I. The two-step approach [8] is described as follows. In the first step,

$$Max\,\beta$$

Subject to (2.3)

$$\beta \le \left[\sum_{j=1}^{n} \frac{\left(c^0_{kj} - \left(c^0_{kj} - c^1_{kj}\right)\alpha\right)x_j - (\tilde{Z}_k)^-_\alpha}{(\tilde{Z}_k)^*_\alpha - (\tilde{Z}_k)^-_\alpha}\right]$$

$$\beta \le \left[\sum_{j=1}^{n} \frac{(\tilde{W}_s)^-_\alpha - \left(c^0_{sj} - \left(c^0_{sj} - c^1_{sj}\right)\alpha\right)x_j}{(\tilde{W}_s)^-_\alpha - (\tilde{W}_s)^*_\alpha}\right]$$

$$\sum_{i=1}^{m}\sum_{j=1}^{n} \frac{[p^0_i + (p^1_i - p^0_i)\alpha].\left[a^0_{ij} + \left(a^1_{ij} - a^0_{ij}\right)\alpha\right]}{B^0 - (B^0 - B^1)\alpha}$$

$$x_j \ge 0,\ i = 1, 2, \ldots, m,\ j = 1, 2, \ldots, n,\ k = 1, 2, \ldots, l,\ \text{and}\ s = 1, 2, \ldots, r$$

The second step model is formed as follows:

$$\text{Max}\,\bar{\beta} = \frac{1}{r + l}\left(\sum_{k=1}^{l}\beta_k + \sum_{s=1}^{r}\beta_s\right)$$

Subject to (2.4)

$$\beta \le \beta_k \le \frac{\sum_{j=1}^{n}\left(c^o_{kj} - \left(c^o_{kj} - c^l_{kj}\right)\alpha.x_j\right) - (\tilde{Z}_k)^-_\alpha}{(\tilde{Z}_k)^*_\alpha - (\tilde{Z}_k)^-_\alpha}$$

$$\beta \le \beta_k \le \frac{(\tilde{W}_s)^-_\alpha - \sum_{j=1}^{n}\left(c^o_{kj} - \left(c^l_{kj} - c^0_{kj}\right)\alpha.x_j\right)}{(\tilde{W}_s)^-_\alpha - (\tilde{W}_s)^*_\alpha}$$

$$\sum_{i=1}^{m}\sum_{j=1}^{n}[p^o_i + (p^l_i - p^o_i)\alpha]\,a_{ij}x_j \le B^0 - (B^0 - B^l)\alpha$$

$$\beta_k,\ \beta_s \in [0, 1],\ x_j \ge 0,\ k = 1, 2, \ldots, l,\ s = 1, 2, \ldots, r,\ \text{and}\ i = 1, 2, \ldots, m$$

Analyzed by Li and Lee [8], in fuzzy environment based on de novo programming, model parameters and investment budget were made more flexible according to the features given below for the product mixture model:

a. When the constraint resources (b_i) were constant, the fuzzy investment budget (B) was determined using the fuzzy unit prices of resources ($\tilde{p}_i$). In an investment budget like this one, the fuzzy investment budget was formed based only on the greatest tolerance value of unit prices.
b. It was estimated that the resource values that were needed to produce each product would decrease based on the smallest tolerance value, and the unit price for the resource amount of each product was evaluated based on the smallest acceptable unit price. In other words, the cost of resource amount was lowered in terms of resources and the lowest acceptable level of unit price of resource amount.
c. Finally, based on the order of goal function, if the goal function was maximization, goal coefficients were increased based on the tolerance value acceptable by the decision maker; and if goal function was minimization, those coefficients were decreased.
d. The model that was formed based on the factors above was solved according to the "$1-\alpha$" risk protection (trust) level. When factors of risky and risk-free production are considered together, it must be determined in the beginning for which factors the decision maker would take risks. Then, the problem is to be modeled this way and solved according to an acceptable "$1-\alpha$" risk protection level.

Rather than modeling the present production models at an acceptable risk level, the model should be formed to produce more efficiently by lowering risks in the beginning of product mixture problems. Based on de novo hypothesis, the increase or decrease of the budget will directly affect production as resource amounts are determined accordingly. Therefore, the increase or decrease of resource unit prices is significant for the budget. Additionally, the definition of resource amounts in a range will reform the budget. Consequently, the appropriate determination of both resource amounts and their unit prices results in a reasonable budget realization.

The risk factors in the studies model and assumptions on their analyses in fuzzy environment are listed as follows:

1. When allocating a budget for investment or resources, the constraint resources should be analyzed in a range based on the de novo programming. The budget is formed based on the constraint resource amounts. As the budget cannot exceed the acceptable tolerance value of constraint resources, there is less risk in terms of the budget, which is formed according to the expectations of the decision maker instead of surpassing. Nevertheless, businesses may only increase their budget up to a certain level.

2. In the fuzzy de novo model proposed by Lee and Li [8], utilization amounts for products were decreased and unit prices were evaluated according to the smallest acceptable value and both of which increased production. On the other hand, the risk factors in production are lowered when the evaluation is based solely on unit prices as resource utilization amounts for each product are fixed. In other words, only the acceptable minimum value of unit prices should be taken into consideration when forming constraints in de novo models.

The proposed approach uses $[f^1, f^0)$ type range for fuzzy coefficients in maximization-type goals and $(f^0, f^1]$ range for fuzzy coefficients in minimization-type goals. $\tilde{a}_{ij}$ coefficient is used as the constant number. The fuzzy coefficients $\tilde{p}_i$ and $\tilde{b}_i$ are of $(f^0, f^1]$ type, and the fuzzy budget $\tilde{B}$ is of $[f^1, f^0)$ type. It means that the optimal system design provides lowest cost, highest profit, lowest resource unit price, and highest budget utilization at α trust level. Consider the fuzzy de novo programming model based on the aforementioned information:

$$\text{Max } \tilde{Z}_k(x) = \sum_{j=1}^{n} \tilde{c}_{kj} x_j$$

$$\text{Min } \tilde{W}_s(x) = \sum_{j=1}^{n} \tilde{c}_{sj} x_j$$

Subject to (2.5)

$$\sum_{j=1}^{n} a_{ij} x_j - b_i \leq 0$$

$$\sum_{i=1}^{m} \tilde{p}_i \tilde{b}_i \otimes \tilde{B}$$

$$x_j \geq 0,\ i = 1, 2, \ldots, m,\ j = 1, 2, \ldots, n,\ k = 1, 2, \ldots, l,\ \text{and}\ s = 1, 2, \ldots, r$$

Solutions are determined as positive ideal or negative ideal for each goal function based on model (2.5):

$$\forall k = 1, 2, \ldots, l$$

$$j_k^* = \left\{ j \in (1, 2, \ldots, n) : \frac{\max\left(\tilde{c}_{kj}\right)_\alpha^U}{\sum_{i=1}^{m} (p)_\alpha^l . a_{ij}} \right\}$$

$$x_k^* = \begin{cases} \dfrac{\left(\tilde{B}\right)_\alpha^U}{\sum_{i=1}^{m} \left(\tilde{p}_i\right)_\alpha^l . a_{ij}}, & j = j_k^* \\ 0, & \text{otherwise} \end{cases} \tag{2.6}$$

$$\left(\tilde{Z}_k\right)^*_\alpha = \sum_{j=1}^{n} \left(\tilde{c}_{kj}\right)^U_\alpha . x^*_k$$

and

$$j^-_k = \left\{ j \in (1, 2, \ldots, n): \frac{\min\left(\tilde{c}_{kj}\right)^L_\alpha}{\sum_{i=1}^{m} (p)^l_\alpha . a_{ij}} \right\}$$

$$x^*_k = \begin{cases} \dfrac{\left(\tilde{B}\right)^U_\alpha}{\sum_{i=1}^{m} \left(\tilde{p}_i\right)^l_\alpha . a_{ij}} & , \quad j = j^*_k \\ 0 & , \quad \text{otherwise} \end{cases} \tag{2.7}$$

$$\left(\tilde{Z}_k\right)^-_\alpha = \sum_{j=1}^{n} \left(\tilde{c}_{kj}\right)^U_\alpha . x^-_k$$

$$\forall s = 1, 2, \ldots, r$$

$$j^*_s = \left\{ j \in (1, 2, \ldots, n): \frac{\min\left(\tilde{c}_{sj}\right)^L_\alpha}{\sum_{i=1}^{m} (p)^l_\alpha . a_{ij}} \right\}$$

$$x^*_s = \begin{cases} \dfrac{\left(\tilde{B}\right)^U_\alpha}{\sum_{i=1}^{m} \left(\tilde{p}_i\right)^l_\alpha . a_{ij}} & , \quad j = j^*_s \\ 0 & , \quad \text{otherwise} \end{cases} \tag{2.8}$$

$$\left(\tilde{W}_s\right)^*_\alpha = \sum_{j=1}^{n} \left(\tilde{c}_{sj}\right)^L_\alpha . x^*_k$$

$$j^-_s = \left\{ j \in (1, 2, \ldots, n): \frac{\max\left(\tilde{c}_{sj}\right)^U_\alpha}{\sum_{i=1}^{m} (p)^l_\alpha . a_{ij}} \right\}$$

$$x^-_s = \begin{cases} \dfrac{\left(\tilde{B}\right)^U_\alpha}{\sum_{i=1}^{m} \left(\tilde{p}_i\right)^l_\alpha . a_{ij}} & , \quad j = j^*_s \\ 0 & , \quad \text{otherwise} \end{cases} \tag{2.9}$$

$$\left(\tilde{W}_s\right)^-_\alpha = \sum_{j=1}^{n} \left(\tilde{c}_{sj}\right)^U_\alpha . x^-_k$$

Based on the "α-cut" value determined by the decision maker, the compromise solution for goal functions is expressed as $\beta = \min\left\{\mu^\alpha_{\tilde{Z}_k}(Z_k), \mu^\alpha_{\tilde{W}_s}(W_s); k = 1, 2, \ldots, l, s = 1, 2, \ldots, r\right\}$. The proposed model for these defined formulations is given below. It makes up the first step of the two-step approach:

$$\text{Max}\,\beta$$

Subject to (2.10)

$$\beta \le \left[\frac{\sum_{j=1}^{n}(\tilde{c}_{kj})_{\alpha}^{U}.x_j - (\tilde{Z}_k)_{\alpha}^{-}}{(\tilde{Z}_k)_{\alpha}^{*} - (\tilde{Z}_k)_{\alpha}^{-}}\right]$$

$$\beta \le \left[\frac{(\tilde{W}_s)_{\alpha}^{-} - \sum_{j=1}^{n}(\tilde{c}_{sj})_{\alpha}^{L}.x_j}{(\tilde{W}_s)_{\alpha}^{-} - (\tilde{W}_s)_{\alpha}^{*}}\right]$$

$$\sum_{i=1}^{m}\sum_{j=1}^{n}(\tilde{p}_i)_{\alpha}^{l}.\, a_{ij}x_j \le (\tilde{B})_{\alpha}^{U}$$

$$x_j \ge 0, \alpha, \beta \in [0; 1]$$

$$j = 1, 2, \ldots, n \text{ and } s = 1, 2, \ldots, r$$

After getting the optimal value of α and β, this solution is tested whether it is efficient or not. If eq. (2.11) is rewritten based on the membership functions of the parameters, the second step of the model is formed as follows:

$$\text{Max}\,\bar{\beta} = \frac{1}{r+l}\left(\sum_{k=1}^{l}\beta_k + \sum_{s=1}^{r}\beta_s\right)$$

Subject to (2.11)

$$\beta \le \beta_k \le \frac{\sum_{j=1}^{n}\left(c_{kj}^{o} - \left(c_{kj}^{o} - c_{kj}^{l}\right)\alpha.x_j\right) - (\tilde{Z}_k)_{\alpha}^{-}}{(\tilde{Z}_k)_{\alpha}^{*} - (\tilde{Z}_k)_{\alpha}^{-}}$$

$$\beta \le \beta_k \le \frac{(\tilde{W}_s)_{\alpha}^{-} - \sum_{j=1}^{n}\left(c_{kj}^{o} - \left(c_{kj}^{l} - c_{kj}^{0}\right)\alpha.x_j\right)}{(\tilde{W}_s)_{\alpha}^{-} - (\tilde{W}_s)_{\alpha}^{*}}$$

$$\sum_{i=1}^{m}\sum_{j=1}^{n}\left[p_i^{o} + (p_i^{l} - p_i^{o})\alpha\right] a_{ij}x_j \le B^{0} - (B^{0} - B^{l})\alpha$$

$$x_j \ge 0, \beta \in [0; 1]$$

$$k = 1, 2, \ldots, l,\ s = 1, 2, \ldots, r, \text{ and } i = 1, 2, \ldots, m$$

The optimal system design is achieved at $\lambda = \alpha = \beta$, where λ represents the minimum level of possibilities under consideration of both the fuzzy goal and the fuzzy coefficients.

2.4 Illustrative example

The multiobjective de novo programming problem prepared by Zeleny [23] was used to compare the results of the proposed approach and the approach by Li and Lee [8]. This problem consists of three maximization-type objective functions, six constraint functions, and three variables as follows:

$$
\begin{aligned}
&\text{Max } Z_1 = 50x_1 + 100x_2 + 17.5x_3 \\
&\text{Max } Z_2 = 92x_1 + 75x_2 + 50x_3 \\
&\text{Max } Z_3 = 25x_1 + 100x_2 + 75x_3
\end{aligned}
$$

Subject to (P1)

$$
\begin{aligned}
&12x_1 + 17x_2 \leq 1400 \\
&3x_1 + 9x_2 + 8x_3 \leq 1,000 \\
&10x_1 + 13x_2 + 15x_3 \leq 1,750 \\
&6x_1 + 16x_3 \leq 1325 \\
&12x_2 + 7x_3 \leq 900 \\
&9.5x_1 + 9.5x_2 + 9.5x_3 \leq 1,075 \\
&x_1, x_2, x_3 \geq 0
\end{aligned}
$$

The resource unit prices for (P1) are \$0.75, \$0.6, \$0.35, \$0.5, \$1.15, \$0.65, respectively. The fuzzy parameter model for (P1) is as follows:

$$
\begin{aligned}
&\text{Max } \tilde{Z}_1 = \widetilde{50}x_1 + 100x_2 + 17.5x_3 \\
&\text{Max } \tilde{Z}_2 = 92x_1 + \widetilde{75}x_2 + 50x_3 \\
&\text{Max } \tilde{Z}_3 = \widetilde{25}x_1 + 100x_2 + 75x_3
\end{aligned}
$$

Subject to (P2)

$$
\begin{aligned}
&12x_1 + 17x_2 \leq \tilde{b}_1 \\
&3x_1 + 9x_2 + 8x_3 \leq 1,000 \\
&10x_1 + 13x_2 + 15x_3 \leq \tilde{b}_3 \\
&6x_1 + 16x_3 \leq 1325 \\
&12x_2 + 7x_3 \leq 900 \\
&9.5x_1 + 9.5x_2 + 9.5x_3 \leq 1,075 \\
&\widetilde{0.75}\tilde{b}_1 + 0.6\tilde{b}_2 + \widetilde{0.35}\tilde{b}_3 + 0.5\tilde{b}_4 + \widetilde{1.15}\tilde{b}_5 + 0.65\tilde{b}_6 \leq \tilde{B} \\
&\quad x_1, x_2, x_3, b_1, b_2, b_3, b_4, b_5, b_6 \geq 0
\end{aligned}
$$

Defined as fuzzy triangular number, goal function coefficients are chosen for each goal function. They are $\widetilde{50} = (40, 50, 80)$, $\widetilde{75} = (70, 75, 90)$, and $\widetilde{25} = (10, 25, 35)$. α – Cut information for the left side and right side corresponding to these coefficients are:

$$\left(\widetilde{50}\right)_{\alpha}^{L} = (40 + 10\alpha), \left(\widetilde{50}\right)_{\alpha}^{R} = (80 - 30\alpha)$$

$$\left(\widetilde{75}\right)_{\alpha}^{L} = (70 + 5\alpha), \quad \left(\widetilde{75}\right)_{\alpha}^{R} = (90 - 15\alpha)$$

$$\left(\widetilde{25}\right)_{\alpha}^{L} = (10 + 15\alpha), \left(\widetilde{25}\right)_{\alpha}^{R} = (35 - 10\alpha)$$

α-Cut information for the left side and the right side are organized as follows for a total of five fuzzy numbers such as two triangular fuzzy numbers accepted for the selected right side constants (b_i), the two trapezoidal fuzzy numbers accepted for the selected unit prices (p_i), and one triangular fuzzy number:

$$\tilde{b}_1 = \widetilde{1400} = (1,300, 1,400, 1,500);$$

$$\left(\widetilde{1400}\right)_{\alpha}^{L} = (1,300 + 100\alpha), \left(\widetilde{14000}\right)_{\alpha}^{R} =, (1500 - 100\alpha)$$

$$\tilde{b}_3 = \widetilde{1750} = (1,500, 1,750, 2,000);$$

$$\left(\widetilde{1750}\right)_{\alpha}^{L} = (1,500 + 100\alpha), \left(\widetilde{1750}\right)_{\alpha}^{R} = (2,000 - 250\alpha)$$

$$\tilde{p}_1 = \widetilde{0.75} = (0.6, 0.7.0.8, 1.15);$$

$$\left(\widetilde{0.75}\right)_{\alpha}^{L} = (0.6 + 0.1\alpha), \left(\widetilde{0.75}\right)_{\alpha}^{R} = (1.15 - 0.35\alpha)$$

$$\tilde{p}_3 = \widetilde{0.35} = (0.3, 0.35, 0.5);$$

$$\left(\widetilde{0.35}\right)_{\alpha}^{L} = (0.3 + 0.05\alpha), \left(\widetilde{0.35}\right)_{\alpha}^{R} = (0.5 - 0.15\alpha)$$

$$\tilde{p}_5 = \widetilde{1.15} = (1, 1.15, 1.2, 1.4);$$

$$\left(\widetilde{1.15}\right)_{\alpha}^{L} = (1 + 0.15\alpha), \left(\widetilde{1.15}\right)_{\alpha}^{R} = (1.4 - 0.2\alpha)$$

Using this formulation, the calculation of the left side and the right side of the budget is as follows. For the left side of the constraint:

$$\left(\widetilde{0.75}\right)_{\alpha}^{L} *(12x_1 + 17x_2) + 0.6*(3x_1 + 9x_2 + 8x_3) + \left(\widetilde{0.35}\right)_{\alpha}^{L} *(10x_1 + 13x_2 + 15x_3)$$

$$+ 0.5*(6x_1 + 16x_3) + \left(\widetilde{1.15}\right)_{\alpha}^{L} *(12x_2 + 7x_3) + 0.65*(9.5x_1 + 9.5x_2 + 9.5x_3$$

or

$$= (0.6 + 0.1\alpha)^*(12x_1 + 17x_2) + 0.6^*(3x_1 + 9x_2 + 8x_3)$$
$$+ (0.3 + 0.05\alpha)^*(10x_1 + 13x_2 + 15x_3)$$
$$+ 0.5^*(6x_1 + 16x_3) + (1 + 0.15\alpha)^*(12x_2 + 7x_3) + 0.65^*(9.5x_1 + 9.5x_2 + 9.5x_3$$

is acquired. The final version of the calculation is

$$(21.175 + 1.7\alpha)x_1 + (37.675 + 4,15\alpha)x_2 + (30.475 + 1,8\alpha)x_3$$

For the right side of the budget constraint:

$$\tilde{B} = (1,500 - 100\alpha)^*(1.15 - 0.35\alpha) + 1,600^*0.6 + (2,000 - 250\alpha)^*(0.5 - 0.15\alpha)$$
$$+ 1,325^*0.5 + 900^*(1.4 - 0.2\alpha) + 1,075^*0.65; \ \tilde{B} = 6,306.25 - 1,245\alpha + 40.5\alpha^2$$

Objective functions are organized as

$$\text{Max}\ \tilde{Z}_1 = (80 - 30\alpha)x_1 + 100x_2 + 17.5x_3$$
$$\text{Max}\ \tilde{Z}_2 = 92x_1 + (90 - 15\alpha)x_2 + 50x_3$$
$$\text{Max}\ \tilde{Z}_3 = (75 - 50\alpha)x_1 + 100x_2 + 75x_3$$

(P1) is formed as below after these processes:

$$\text{Max}\ \tilde{Z}_1 = (80 - 30\alpha)x_1 + 100x_2 + 17.5x_3$$
$$\text{Max}\ \tilde{Z}_2 = 92x_1 + (90 - 15\alpha)x_2 + 50x_3$$
$$\text{Max}\ \tilde{Z}_3 = (75 - 50\alpha)x_1 + 100x_2 + 75x_3$$

Subject to (P3)

$$(21.175 + 1.7\alpha)x_1 + (37.675 + 4,15\alpha)x_2 + (30.475 + 1,8\alpha)x_3$$
$$\leq 6,306.25 - 1,245\alpha + 40.5\alpha^2$$
$$x_1, x_2, x_3 \geq 0, \alpha \in [0, 1]$$

First, if a parametric solution is made for $\alpha = 1$ (P3), it is obtained as follows:

$$\text{Max}\ Z_1 = 50x_1 + 100x_2 + 17.5x_3$$
$$\text{Max}\ Z_2 = 92x_1 + 75x_2 + 50x_3$$
$$\text{Max}\ Z_3 = 25x_1 + 100x_2 + 75x_3$$

(P3.1)

Subject to

$$22.875x_1 + 41.825x_2 + 32.275x_3 \leq 5,101.75$$
$$x_1, x_2, x_3 \geq 0$$

acquired.

Positive ideal solutions for each objective function from (P3.1) are calculated as

$$\left(\tilde{Z}_1\right)^*_{\alpha=1} = 13,528.99;\ \left(\tilde{Z}_2\right)^*_{\alpha=1} = 22,757.68;\ \left(\tilde{Z}_3\right)^*_{\alpha=1} = 14,621.45$$

and negative ideal solutions are calculated as

$$\left(\tilde{Z}_1\right)^-_{\alpha=1} = 0;\ \left(\tilde{Z}_2\right)^-_{\alpha=1} = 0;\ \left(\tilde{Z}_3\right)^-_{\alpha=1} = 0$$

The first phase is written for (P3.1) using eq. (2.10) as follows:

$$\text{Max}\,\beta$$

Subject to (P3.2)

$$\beta \le \left[\frac{50x_1 + 100x_2 + 17.5x_3 - 0}{13,528.995 - 0}\right]$$

$$\beta \le \left[\frac{92x_1 + 75x_2 + 50x_3 - 0}{22,757.68 - 0}\right]$$

$$\beta \le \left[\frac{25x_1 + 100x_2 + 75x_3 - 0}{14,621.45 - 0}\right]$$

$$22.875x_1 + 41.825x_2 + 32.275x_3 \le 5,101.75$$

$$x_1, x_2, x_3 \ge 0, \beta \in [0;1]$$

Values obtained from the solution of (P3.2) are given in Table 2.1.

Table 2.1: Decision variables of Phase I for (P3.1).

Decision variables	Z_3	Z_3	Z_3
x_1	101.208	101.208	101.208
x_2	66.625	66.625	66.625
x_3	0	0	0
Objective function value	11,722.9	14,308.01	10,457.8

Information in Table 2.1 is the result of the solution of (P3.1)'s first phase. The objective function value of (P3.1) is $\beta = 0.6287$ that is for $\alpha = 1$, $\beta = 0.6287$. According to this result, the solution of (P3) will take place in the range of [0.6287; 1]. Because of $\alpha \neq \beta$, the solution should be continued by selecting a new α. The result obtained for the selected $\alpha = 0.71999$ is optimal. First, if (P3) is taken as $\alpha = 0.71999$

$$\begin{aligned}&\text{Max } Z_1 = 58.4x_1 + 100x_2 + 17.5x_3\\&\text{Max } Z_2 = 92x_1 + 79.2x_2 + 50x_3\\&\text{Max } Z_3 = 27.8x_1 + 100x_2 + 75x_3\end{aligned} \tag{P3.3}$$

Subject to

$$\begin{aligned}&22.399x_1 + 40.633x_2 + 28.521x_3 \le 5,430.845\\&x_1, x_2, x_3 \ge 0\end{aligned}$$

is obtained. First, positive and negative ideal solutions for (P3.3) are given as follows:

$$(\tilde{Z}_1)^*_{\alpha=0.71999} = 14,159.62;\ (\tilde{Z}_2)^*_{\alpha=0.71999} = 22,306.25;\ (\tilde{Z}_3)^*_{\alpha=0.71999} = 14,281.17$$

and

$$(\tilde{Z}_1)^-_{\alpha=0.71999} = 0;\ (\tilde{Z}_2)^-_{\alpha=0.71999} = 0;\ (\tilde{Z}_3)^-_{\alpha=0.71999} = 0$$

Phase I is arranged using these solutions as follows:

Phase I

$$\text{Max}\,\beta$$

Subject to

$$\begin{aligned}&58.4x_1 + 100x_2 + 17.5x_3 - 14,159.62\beta \ge 0\\&92x_1 + 79.2x_2 + 50x_3 - 22,306.25\beta \ge 0\\&27.8x_1 + 100x_2 + 75x_3 - 14,281.17\beta \ge 0\end{aligned}$$

Subject to (P3.4)

$$\begin{aligned}&22.399x_1 + 40.633x_2 + 28.521x_3 \le 5,430.845\\&x_1, x_2, x_3 \ge 0, \beta \in [0; 1]\end{aligned}$$

$\beta = 0.71999$, $x_1 = 113.057869$, $x_2 = 71.332710$, $x_3 = 0$ are obtained from (P3.4)' solution.

In the first phase, $\beta = 0.71999$ is obtained for $\alpha = 0.71999$. It is now tested in phase 2 whether the solution of (P3.4) is efficient:

$$\text{Max}\,\bar{\beta} = \frac{1}{3}(\beta_1 + \beta_2 + \beta_2)$$

Subject to (P3.5)

$$\beta \le \beta_1 \le \left[\frac{58.4x_1 + 100x_2 + 17.5x_3 - 0}{14,159.62 - 0}\right]$$

$$\beta \le \beta_2 \le \left[\frac{92x_1 + 79.2x_2 + 50x_3 - 0}{22,306.25 - 0}\right]$$

$$\beta \le \beta_3 \le \left[\frac{27.8x_1 + 100x_2 + 75x_3 - 0}{14,281.17 - 0}\right]$$

$$22.399x_1 + 40.633x_2 + 28.521x_3 \le 5,430.845$$

$$\beta_1 \ge 0.71999$$

$$\beta_2 \ge 0.71999$$

$$\beta_3 \ge 0.71999$$

$$x_1, x_2, x_3 \ge 0, \beta \in [0;1]$$

Variable values obtained from (P3.5)'s solution created for phase 2 are specified as

$$\bar{\beta} = 0.80334,\ \beta_1 = 0.970072,\ \beta_2 = 0.71999,\ \beta_3 = 0.71999$$

$$x_1 = 113.057861,\ x_2 = 71.332710,\ \text{and}\ x_3 = 0$$

This result shows that values obtained from phase 1 are efficient. Depending on the different values of α, variable and objective function values are given in Table 2.2.

Table 2.2: The results of (P3).

α	β	Z_1	Z_2	Z_3	**Budget**
1.0	0.6973268	13002,15	15869.4	11598.99	5101.75
0.75	0.7169044	12664.65	15584.99	11268.66	5395.281
0.73	0.7191670	12743.9	15673.59	11341.23	5418,982
0.725	0.7188874	12761.55	15718.16	11351,69	5424.913
0.71999	*0.71999*	*12786.05*	*15751.14*	*11372,84*	*5430.845*

The optimal result is obtained at $\alpha = \beta = 0.71999$. At this point, resource utilization amounts according to decision variable values are 2569.328, 981.159, 2,057.886, 678.342, 855.984, and 1,751.696, respectively.

2.5 Conclusions

As the main objective in traditional production processes is maximum profit and minimum production process, conciliatory or satisfactory solution is sought instead of optimal solution due to the multiple objective functions. In multiobjective de novo programming problems, it is possible to specify the optimal solutions at the end of similar processes. For example, if multiobjective de novo programming problem is solved through compromise programming, the result can be named as compromise optimal system design, or if it is solved through goal programming, the result can be named as satisfactory optimal system design. In this study, multiobjective de novo programming problem is solved because the solution of this problem is made in terms of fuzzy set theory, and is named as fuzzy optimal system design. However, the concept of compromise solution is used in some studies instead of fuzzy solution.

In this study, the model offered for fuzzy optimal system design comes into prominence with some features. First of them is although it is possible to define a_{ij} parameters as a fuzzy number, it is wrong to fuzzify this parameter in the production process. This is because the change in the parameter a_{ij} causes the amount used to produce the product during the production process and this disrupts the generated model. Also, p_i and b_i fuzzy parameters were used while setting a fuzzy budget. This situation caused the upper limit of the fuzzy b_i determined by the decision maker to directly affect the budget while creating the fuzzy budget. The proposed approach can be accepted as an extension of Li and Lee's [8] approach because the solution procedure of the proposed approach is the same as that of Li and Lee's approach. Although it is possible to compare the results of Li and Lee's [8] approach with the result of the proposed approach, it is not recommended because of the differences explained earlier. Fuzzy parameters are different from each other, even though the solution of both approaches is the same.

References

[1] Li R.J. Multiple objective decision-making in a fuzzy environment, Yayınlanmamış Doktora Tezi, Kansas State University, 1990.
[2] Eiselt H.A. & Sandblom C.L. Linear programming and its applications, Springer-Verlag Berlin Heidelberg, 2007.
[3] Yu P.L. Introduction to domination structures in multicriteria problems, In: Cochrane J.L. & Zeleny M., eds., Multiple criteria decision making, University of South Carolina Press, Columbia SC, 1973.
[4] Mollaghasemi M. & Pet-Edwards J. Technical briefing: making multiple- objective decisions, IEEE Computer Society Press Los Alamitos, California, 1997.
[5] Zimmermann H.J. (1978) Fuzzy programming and linear programming with several functions, Fuzzy Sets and Sysytems, 1, 45–55.

[6] Zeleny M. Multiobjective Design of High-Productivity Systems, Proceedings of Joint. Automatic Control Conference, July 27–30, Purdue University.1976.
[7] Zeleny M. (1981) On the squanering of resources and profits via linear programming, Interfaces, 11(5), 101–107.
[8] Li R.J. & Lee E.S. (1993) Fuzzy multiple objective programming and compromise programming with pareto optimum, Fuzzy Sets and Systems, 53, 275–288.
[9] Zeleny M. Systems Approach to Multiple Criteria Decision Making: Metaoptimum, In: Toward interactive and intelligent decision support systems, edited by, Sawaragi Y., Inoue K., & Nakayama H., Springer-Verlag, New York, 1987, 28–37.
[10] Zeleny M. (1990) Optimizing given systems vs. designing optimal systems: the De Novo programming approach, International Journal of General Systems, 17(4), 295–307.
[11] Zeleny M. Multicriterion Design of High-Productivity Systems: Extensions and Applications, In: Decision making with multiple objectives, edited by, Haimes Y.Y. & Chankong V., Springer-Verlag, New York, 1985, 308–321.
[12] Hessel M. & Zeleny M. (1987) Optimal system design: towards new interpretation of shadow prices ın linear programming, Computer Operational Research, 14(4), 265–271.
[13] Shi Y. (1995) Studuies on optimum-path ratios in multicriteria De Novo programming problems, Computers Mathamatical Applications, 29(5), 43–50.
[14] Umarusman N. (2013) Min-max goal programming approach for solving multi-objective de novo programming problems, International Journal of Operations Research, 10(2), 92–99.
[15] Zhuang Z.Y. & Hocine A. (2018) Meta goal programing approach for solving multi-criteria de Novo programming problem, European Journal of Operational Research, 265(1), 228–238.
[16] Lai Y.-J. & Hwang C.-L. Fuzzy mathematical programming; methods and applications (First edition), Berlin Heidelberg, Springer-Verlag, 1992.
[17] Li R.J. & Lee E.S. (1990a) Approaches to multicriteria de novo programs, Journal of Mathematical Analysis and Applications, 153, 97–111.
[18] Li R.J. & Lee E.S. (1990b) Multi-criteria de novo programming with fuzzy parameters, Computers Mathamatical Applications, 19(56), 13–20.
[19] Bhattacharya D. & Chakraborty S. (2018) Solution of the general multi-objective De-Novo programming problem using compensatory operator under fuzzy environment, Journal of Physics: Conference Series, 1039(1), 1–8.
[20] Umarusman N. (2018) A fuzzy approach proposal in the solution of multi objective De Novo programming problems and a business application, Business and Economics Research Journal, 9(4), 825–838.
[21] Zeleny M. Linear multiobjective programming, Springer-Verlag, New York, 1974.
[22] Zadeh L.A. (1965) Fuzzy Sets, Information and Control, 8, 338–353.
[23] Zeleny M. (1986) An external reconstruction approach (ERA) to linear programming, Computers and Operations Research, 13(1), 95–100.
[24] Babic Z. & Pavic I. (1999) Multicriterial production planning by De Novo programming approach, International Journal of Production Economics, 43, 59–66.
[25] Tabucanon M.T. Multiple criteria decision making in industry, Elsevier, New York, 1998.
[26] Bhattacharya D. & Banik S. (2019) Hacettepe University Bulletin of Natural Sciences and Engineering Series B: Mathematics and Statistics, 49(1), 1–14.

Sumit Kumar Maiti, Sankar Kumar Roy

3 A probabilistic two-level programming in noncooperative game under fuzzy environment: comparative study

Abstract: This chapter develops a computational algorithm to solve a stochastic two-level programming in a noncooperative game using a fuzzy optimization technique. It also includes all parameters such as fuzzy variables except the constraints on the right side that follow an extreme value distribution. Mellin transformation is considered to convert the fuzzy numbers into the crisp value. Based on stochastic programming, probabilistic constraints are converted into the deterministic form. Furthermore, as the fuzzy programming is used for the degree of acceptance, we present a comparative study of nonlinear and linear membership functions for the acceptance corresponding to LINGO 15.0 iterative scheme and genetic algorithm to see its impact on optimization. A numerical experiment is considered in order to present the applicability and feasibility of the proposed algorithm. Finally, conclusion about the findings and outlook are presented.

Keywords: Bilevel programming, fuzzy number, GA, membership function, Stackelberg game, stochastic programming, satisfactory solution

3.1 Introduction

Two-level programming problem (TLPP) consists of two levels in any hierarchical decision process, and large numbers of research works have been investigated on TLPP in the past few decades ([19],[20],[24]). In TLPP, each decision maker (DM) optimizes his/her own objectives without considering the other one. Stackelberg game was first introduced by an economist H. von Stackelberg to solve noncooperative decision problems. In Stackelberg game, the leader has the capability to enforce his/her decision on the followers. Many researchers have carried out investigations on the Stackelberg game (cf [1–4]) Ref. [17]. In real-life-based applications, when the input data of any mathematical model are inexistent or scarce, data are considered as uncertain. To tackle the uncertainty problem, two principal resources say, fuzziness and randomness, are considered. Fuzzy programming (cf

Sumit Kumar Maiti, School of Applied Sciences and Humanities, Haldia Institute of Technology, West Bengal, India
Sankar Kumar Roy, Department of Applied Mathematics with Oceanology and Computer Programming, Vidyasagar University, West Bengal, India

https://doi.org/10.1515/9783110628616-003

[5, 6]) developed by characterizing the input data as fuzzy sets, and stochastic programming is also developed to express the input data as random variables. Using the Mellin transformation, all fuzzy numbers (triangular or trapezoidal) are reduced to a crisp form. The constraint parameters on the right side follow an extreme value distribution Ref. [22]. Then, using a stochastic programming, these constraints are converted to the deterministic form and the equivalent crisp problems are solved by LINGO 15.0 iterative scheme and the results are compared. This chapter includes the following:
- Mellin transformation is used for the fuzzy numbers (triangular or trapezoidal) into crisp value.
- The constraint parameters on the right side are considered to be an extreme value distribution.
- A computational algorithm is developed to solve a noncooperative game.

We develop a computational algorithm for solving a stochastic TLPP in a noncooperative game involving an extreme value distribution by fuzzy programming. Our aim is to study the impacts of different types of membership functions. For that instance, we consider a linear function and a nonlinear function. The research works of several authors related to this area are shown in Table 3.1.

Table 3.1: Research works of several authors related to this area.

Author(s)	Real number	Fuzzy number	Fuzzy random	Multichoice
Anandalingam and Apprey [1]	Yes			
Anandalingam and White [7]				Yes
Hamidi and Mishmast [8]		Yes		
Maity and Roy [2]		Yes		Yes
Roy and Maiti [4]			Yes	Yes
Sakawa and Matsui [9]			Yes	
Singh and Chakraborty [10]			Yes	
Yeh et al. [11]			Yes	
Youness et al. [12]		Yes		
Zheng et al. [13]	Yes			
Our chapter		Yes	Yes	

The sequence of the chapter is given as follows: Section 3.2 introduces the basic knowledge of fuzzy set. In Section 3.3, model formulation of stochastic TLPP in a noncooperative game involving an extreme value distribution is presented. The computation algorithm of fuzzy optimization is discussed in Section 3.4. A numerical example is presented in Section 3.5. Results are reflected in Section 3.6. The conclusion and an outlook of future research directions are described in Section 3.7.

3.2 Preliminaries

Here, we recall the useful definitions of fuzzy set and related theorems that are to be required for our subsequent developments.

Definition 3.1 ([11]). Let $\tilde{A}$ be a fuzzy set defined on X, the universal set. Then $\tilde{A} = \{(x, \mu_{\tilde{A}}(x)) : x \in X\}$, where $\mu_{\tilde{A}} : X \to [0,1]$ is the membership function such that $\mu_{\tilde{A}}(x) = 0$ if $x \notin \tilde{A}$, $\mu_{\tilde{A}}(x) = 1$ if $x \in \tilde{A}$.

Definition 3.2 ([18]). Let $\tilde{A}$ be a fuzzy set defined on $\mathbb{R}$, the set of real numbers, which is said to be a fuzzy number if its membership function satisfies the following conditions:
(a) $\mu_{\tilde{A}}(x) : \mathbb{R} \to [0,1]$ is continuous,
(b) $\mu_{\tilde{A}}(x) = 0 \quad \forall x \in (-\infty, c] \cup [b, \infty)$,
(c) $\mu_{\tilde{A}}(x)$ is strictly increasing on $[c,d]$ and strictly decreasing on $[a,b]$, $(c < d < a < b)$,
(d) $\mu_{\tilde{A}}(x) = \rho \quad \forall x \in [d,a]$, where $0 < \rho \le 1$.

Definition 3.3 ([14]). Let $\tilde{B} = (b_1, b_2, b_3)$ be a fuzzy number. Then $\tilde{B}$ is said to be a triangular fuzzy number if $\mu_{\tilde{B}}(x)$ is given by

$$\mu_{\tilde{B}}(x) = \begin{cases} \left(\frac{x - b_1}{b_2 - b_1}\right), & \text{if } b_1 \le x \le b_2 \\ \left(\frac{b_3 - x}{b_3 - b_2}\right), & \text{if } b_2 \le x \le b_3 \\ 0, & \text{if otherwise} \end{cases} \tag{3.1}$$

Definition 3.4 ([14]). Let $\tilde{C} = (c_1, c_2, c_3, c_4)$ be a fuzzy number. Then $\tilde{C}$ is said to be a trapezoidal fuzzy number if $\mu_{\tilde{C}}(x)$ is given by

$$\mu_{\tilde{C}}(x) = \begin{cases} \left(\frac{x - c_1}{c_2 - c_1}\right), & \text{if } c_1 \le x \le c_2 \\ 1, & \text{if } c_2 \le x \le c_3 \\ \left(\frac{c_4 - x}{c_4 - c_3}\right), & \text{if } c_3 \le x \le c_4 \\ 0, & \text{if otherwise} \end{cases} \tag{3.2}$$

Definition 3.5 ([15]). Let $g(x)$ be a probability density function (pdf). Then Mellin transform $M_X(r)$ of $g(x)$ is given by

$$M_X(r) = \int_0^\infty x^{r-1} g(x)\, dx$$

provided the integral exists.

3.3 Model formulation

This chapter discusses on a stochastic TLPP for a noncooperative game in which cost coefficients of both level objectives are either triangular or trapezoidal fuzzy numbers. Generally, the constraint coefficients in a TLPP are considered as crisp. But in real-life situations, the constraint parameters are imprecise. So, we consider that the constraint parameters are of two different natures. The left side of the constraints is a fuzzy number, whereas the right side follows an extreme value distribution. The following assumptions are considered to design the model:

- $\tilde{c}_{1j}$: fuzzy number (triangular or trapezoidal) for leader objectives, $j = 1, 2, \ldots, n$;
- $\tilde{c}_{2j}$: fuzzy number (triangular or trapezoidal) for follower objectives, $j = 1, 2, \ldots, n$;
- $\tilde{a}_{ij}$: a $m \times n$ matrix with fuzzy numbers;
- $\bar{b}_i$: random variable follows an extreme value distribution.

Therefore, the model on a stochastic TLPP for a noncooperative game involving extreme value distribution is defined as follows:

P1

$$\begin{aligned} \text{maximize} \quad & Z_1(x) = \sum_{j=1}^{n} \tilde{c}_{1j} x_j \\ \text{maximize} \quad & Z_2(x) = \sum_{j=1}^{n} \tilde{c}_{2j} x_j \\ \text{subject to} \quad & \text{pr}\left(\sum_{j=1}^{n} \tilde{A}_{ij} x_j \le \bar{B}_i \right) \ge 1 - \gamma_i, \quad i = 1, 2, \ldots, m \\ & x_j \ge 0 \quad \forall j \ \text{ and } \ 0 < \gamma_i < 1 \quad \forall i \end{aligned} \tag{3.3}$$

3.3.1 Deterministic form of a fuzzy number

Here, we present a defuzzification procedure of a triangular or trapezoidal fuzzy number.

3.3.1.1 Triangular fuzzy number

Let $\tilde{a} = (a_1, a_2, a_3)$ be the triangular fuzzy number. Then the crisp value is obtained by finding the expected value using the pdf corresponding to the membership function.

Therefore, the proportional pdf corresponding to $\tilde{a}$ is defined as follows: $f_{\tilde{a}}(x) = d\mu_{\tilde{a}}(x)$, where $\mu_{\tilde{a}}(x)$ is given by eq. (3.1) .

To find d such that $\int_{-\infty}^{\infty} f_{\tilde{a}}(x)\,dx = 1$, we get $d = \frac{2}{a_3 - a_1}$.

Then the proportional pdf corresponding to $\tilde{a}$ is given by

$$f_{\tilde{a}}(x) = \begin{cases} \frac{2(x - a_1)}{(a_2 - a_1)(a_3 - a_1)}, & \text{if } a_1 \le x \le a_2 \\ \frac{2(a_3 - x)}{(a_3 - a_2)(a_3 - a_1)}, & \text{if } a_2 \le x \le a_3 \\ 0, & \text{if otherwise} \end{cases}$$

According to Definition 3.5, we obtain

$$M_{\tilde{a}}(r) = \int_0^{\infty} x^{r-1} f_{\tilde{a}}(x)\,dx = \int_{a_1}^{a_2} x^{r-1} \frac{2(x - a_1)}{(a_2 - a_1)(a_3 - a_1)} dx + \int_{a_2}^{a_3} x^{r-1} \frac{2(a_3 - x)}{(a_3 - a_2)(a_3 - a_1)} dx$$

On integration, we obtain

$$M_{\tilde{a}}(r) = \frac{2}{(a_3 - a_1) r (r+1)} \left[\frac{a_3 (a_3^r - a_2^r)}{a_3 - a_2} - \frac{a_1 (a_2^r - a_1^r)}{a_2 - a_1} \right]$$

Therefore, the mean ($E[\tilde{a}]$) and variance ($\sigma_{\tilde{a}}^2$) corresponding to $\tilde{a}$ are obtained as follows:

$$E[\tilde{a}] = M_{\tilde{a}}(2) = \frac{a_1 + a_2 + a_3}{3}$$

$$\sigma_{\tilde{a}}^2 = M_{\tilde{a}}(3) - [M_{\tilde{a}}(2)]^2 = \frac{a_1^2 + a_2^2 + a_3^2 - a_1 a_2 - a_2 a_3 - a_3 a_1}{18}$$

Then the crisp value of $\tilde{a}$ is given by

$$\text{Crisp}(\tilde{a}) = a' = \frac{a_1 + a_2 + a_3}{3} \tag{3.4}$$

3.3.1.2 Trapezoidal fuzzy number

Let $\tilde{b} = (b_1, b_2, b_3)$ be the trapezoidal fuzzy number, and the proportional pdf corresponding to $\tilde{b}$ is defined as follows: $f_{\tilde{b}}(x) = c\mu_{\tilde{b}}(x)$, where $\mu_{\tilde{b}}(x)$ is given by eq. (3.2) and c is given by relation

$$\int_{-\infty}^{\infty} f_{\tilde{b}}(x)\,dx = 1 \tag{3.5}$$

From eq. (3.5), we obtain $c = \frac{2}{b_4 + b_3 - b_2 - b_1}$.

Therefore, the proportional pdf corresponding to $\tilde{B}$ is given by

$$f_{\tilde{b}}(x) = \begin{cases} \frac{2(x - b_1)}{(b_2 - b_1)(b_4 + b_3 - b_2 - b_1)}, & \text{if } b_1 \le x \le b_2 \\ \frac{2}{(b_4 + b_3 - b_2 - b_1)}, & \text{if } b_2 \le x \le b_3 \\ \frac{2(b_4 - x)}{(b_4 - b_3)(b_4 + b_3 - b_2 - b_1)}, & \text{if } b_3 \le x \le b_4 \\ 0, & \text{if otherwise} \end{cases}$$

According to Definition 3.5, we obtain

$$\begin{aligned} M_{\tilde{b}}(r) &= \int_{0}^{\infty} x^{r-1} f_{\tilde{b}}(x)\,dx \\ &= \int_{b_1}^{b_2} x^{t-1} \frac{2(x - b_1)}{(b_2 - b_1)(b_4 + b_3 - b_2 - b_1)} dx + \int_{b_2}^{b_3} x^{r-1} \frac{2}{(b_4 + b_3 - b_2 - b_1)} dx \\ &\quad + \int_{b_3}^{b_4} x^{r-1} \frac{2(b_4 - x)}{(b_4 - b_3)(b_4 + b_3 - b_2 - b_1)} dx \end{aligned}$$

On integration, we obtain

$$M_{\tilde{b}}(r) = \frac{2}{(b_4 + b_3 - b_2 - b_1) r(r+1)} \left[\frac{\left(b_4^{r+1} - b_3^{r+1}\right)}{b_4 - b_3} - \frac{\left(b_2^{r+1} - b_1^{r+1}\right)}{b_2 - b_1} \right]$$

Then, the mean $\left(E\left[\tilde{b}\right]\right)$ and variance $\left(\sigma_{\tilde{b}}^2\right)$ corresponding to $\tilde{b}$ are obtained as follows:

$$E\left[\tilde{b}\right] = M_{\tilde{b}}(2) = \frac{1}{3}\left[(b_1 + b_2 + b_3 + b_4) + \frac{(b_1 b_2 - b_3 b_4)}{(b_4 + b_3 - b_2 - b_1)}\right]$$

$$\sigma_{\tilde{b}}^2 = M_{\tilde{b}}(3) - \left[M_{\tilde{b}}(2)\right]^2 = \frac{1}{6}\left[\left(b_1^2 + b_2^2 + b_3^2 + b_4^2\right) + \frac{(b_1 + b_2)\left(b_3^2 + b_4^2\right) - (b_3 + b_4)\left(b_1^2 + b_2^2\right)}{(b_4 + b_3 - b_2 - b_1)}\right]$$

Then the crisp value of $\tilde{b}$ is given by

$$\text{Crisp}\left(\tilde{b}\right) = b' = \frac{1}{3}\left[(b_1 + b_2 + b_3 + b_4) + \frac{(b_1 b_2 - b_3 b_4)}{(b_4 + b_3 - b_2 - b_1)}\right] \tag{3.6}$$

Using eq. (3.4) (or eq. (3.6)), model 1 is transformed into the following model with the probabilistic constraint as follows:

P2

$$\begin{aligned}
&\text{maximize} \quad Z_1(x) = \sum_{j=1}^{n} c'_{1j}x_j \\
&\text{maximize} \quad Z_2(x) = \sum_{j=1}^{n} c'_{2j}x_j \\
&\text{subject to} \quad \text{pr}\left(\sum_{j=1}^{n} \tilde{A}_{ij}x_j \le \overline{B_i}\right) \ge 1-\gamma_i, \quad i=1,2,\ldots,m \\
&\qquad\qquad x_j \ge 0 \ \ \forall j \ \text{ and } \ 0<\gamma_i<1 \ \ \forall i
\end{aligned}$$

3.3.2 Deterministic form of probabilistic constraint

Here, $\overline{B_i}$ $\forall i$ are independent random variables that follow an extreme value distribution with δ_i and β_i as location and scale parameters, respectively, where the aspiration level is γ_i, $0<\gamma_i<1$.

Now, the pdf of $\overline{B_i}$ $\forall i$ is defined as follows:

$$f(\overline{B_i}) = \frac{1}{\beta_i}\exp\left(-\left(\frac{\overline{B_i}-\delta_i}{\beta_i}\right)\right)\exp\left(-\left(\exp\left(-\left(\frac{\overline{B_i}-\delta_i}{\beta_i}\right)\right)\right)\right), \quad -\infty<\overline{B_i}<\infty \ \text{ and } \ \delta_i>0,\ \beta_i>0 \ \ \forall i \tag{3.7}$$

Therefore, eq. (3.3) can be written as follows:

$$\int_{\sum_{j=1}^{n} A'_{ij}x_j}^{\infty} f(\overline{B_i})\, d\overline{B_i} \ge 1-\gamma_i \tag{3.8}$$

Using eq. (3.7), the above eq. (3.8) can be written as

$$\int_{\sum_{j=1}^{n} A'_{ij}x_j}^{\infty} \frac{1}{\beta_i}\exp\left(-\left(\frac{\overline{B_i}-\delta_i}{\beta_i}\right)\right)\exp\left(-\left(\exp\left(-\left(\frac{\overline{B_i}-\delta_i}{\beta_i}\right)\right)\right)\right) d\overline{B_i} \ge 1-\gamma_i \tag{3.9}$$

Let us take $\exp\left(-\left(\frac{\overline{B_i}-\delta_i}{\beta_i}\right)\right) = q$. Then eq. (3.9) reduces to $\int_0^{\exp\left(-\left(\frac{\sum_{j=1}^{n} A'_{ij}x_j-\delta_i}{\beta_i}\right)\right)} e^{-q}\, dq \ge 1-\gamma_i$.

On integration, we get

$$\exp\left(-\exp\left(-\left(\frac{\sum_{j=1}^{n} A'_{ij}x_j - \delta_i}{\beta_i}\right)\right)\right) \leq 1 - \gamma_i$$

Taking logarithm of both sides two times and rearranging, we get

$$\sum_{j=1}^{n} A'_{ij}x_j \leq \delta_i - \beta_i\left[\ln\left(-\ln(\gamma_i)\right)\right] \tag{3.10}$$

Then we obtained the following deterministic model:

P3

$$\begin{aligned} &\text{maximize} \quad Z_1(x) = \sum_{j=1}^{n} c'_{1j}x_j \\ &\text{maximize} \quad Z_2(x) = \sum_{j=1}^{n} c'_{2j}x_j \\ &\text{subject to} \quad \sum_{j=1}^{n} A'_{ij}x_j \leq \delta_i - \beta_i\left[\ln\left(-\ln(\gamma_i)\right)\right] \\ &\qquad\qquad\quad x_j \geq 0 \ \ \forall j \ \text{ and } \ 0 < \gamma_i < 1, \ \delta_i > 0, \ \beta_i > 0 \ \ \forall\, i \end{aligned}$$

3.4 Computational algorithm

After getting the deterministic form of P1, we develop the following two algorithms corresponding to the satisfactory solution using fuzzy approach. The following steps of fuzzy programming are given:

Algorithm 3.1 (For linear membership function).

Step 4.1: Independently solve the leader and follower objectives with respect to the constraints and assign the best (Z_s^0) and worst (Z_s^1) solutions for each objective function, $s = 1, 2$.

Step 4.2: A linear membership function is considered for each objective function as follows:

$$\mu_s(Z_s(x)) = \begin{cases} 1, & \text{if } Z_s(x) Z_s^0 \\ \left(\frac{Z_s(x) - Z_s^1}{Z_s^0 - Z_s^1}\right) & \text{if } Z_s^1 \leq Z_s(x) \leq Z_s^0 0, \\ 0, & \text{if } Z_s(x) \leq Z_s^1, \end{cases}$$

Again, considering a linear membership function for the upper-level decision variable (x_1) with f tolerance value is defined by

$$\mu_{x_1}(x_1) = \begin{cases} \left(\frac{x_1 - x_1^t + f}{f}\right), & \text{if } x_1^t - f \le x_1 \le x_1^t \\ \left(\frac{x_1^t + f - x_1}{f}\right), & \text{if } x_1^t \le x_1 \le x_1^t + f \\ 0, & \text{if otherwise} \end{cases}$$

where x_1^t is the solution set of the top-level DM.

Step 4.3: According to Zimmerman approach [16], solve the following problem:

P4

$$\begin{aligned} &\text{maximize} && \lambda \\ &\text{subject to} && \mu_s(Z_s(x)) \ge \lambda \quad \forall s \\ &&& \mu_{x_1}(x_1) \ge \lambda, \\ &&& \text{eq. (3.10)}, \\ &&& x \ge 0 \text{ and } \lambda \in [0,1] \end{aligned}$$

If the top-level DM is satisfied with the obtained optimal result, then the solutions become a satisfactory solution and go to Step 4.4. Otherwise, top-level DM updated his/her membership function and then go to Step 4.2.

Step 4.4: Stop.

Algorithm 3.2 (For nonlinear membership function).
Repeat Step 4.1 and follow the steps.

Step 4.2: Consider a nonlinear membership for each objective, and the top-level decision variable (x_1) with tolerance value f is given by

$$\mu_s(Z_s(x)) = \begin{cases} 1, & \text{if } Z_s(x) > Z_s^0 \\ \left(\frac{e^{-r\left(\frac{Z_s(x) - Z_s^1}{Z_s^0 - Z_s^1}\right)} - e^{-r}}{1 - e^{-r}}\right), & \text{if } Z_s^1 \le Z_s(x) \le Z_s^0 \\ 0, & \text{if } Z_s(x) \le Z_s^1 \end{cases}$$

$$\mu_{x_1}(x_1) = \begin{cases} \frac{e^{-r\left(\frac{x_1 - x_1^t + f}{f}\right)} - e^{-r}}{1 - e^{-r}}, & \text{if } x_1^t - f \le x_1 \le x_1^t \\ \frac{e^{-r\left(\frac{x_1^t - x_1 + f}{f}\right)} - e^{-r}}{1 - e^{-r}}, & \text{if } x_1^t \le x_1 \le x_1^t + f \\ 0, & \text{if otherwise} \end{cases}$$

where x_1^t is the solution of the top-level DM. After that repeat Step 4.3 and Step 4.4.

3.5 Numerical experiment

Here, we present two numerical examples corresponding to triangular and trapezoidal fuzzy numbers and one practical application.

3.5.1 Example 1

Let us consider the following example:

P5

$$\begin{aligned}
\text{maximize}\quad & Z_1(x) = (200,302,388)x_1 + (234,280,347)x_2 + (187,199,223)x_3 + (207,289,301)x_4\\
\text{maximize}\quad & Z_2(x) = (89,107,112)x_1 + (106,127,132)x_2 + (67,80,93)x_3 + (73,79,104)x_4\\
\text{subject to}\quad & (34,42,49)x_1 + (27,37,52)x_2 + (30,47,53)x_3 + (18,22,28)x_4 \le b_1\\
& (32,37,42)x_1 + (18,22,25)x_2 + (25,27,29)x_3 + (30,32,36)x_4 \le b_2\\
& (26,36,42)x_1 + (16,20,27)x_2 + (41,44,49)x_3 + (51,55,59)x_4 \ge b_3\\
& (38,43,47)x_1 + (52,56,57)x_2 + (61,65,68)x_3 + (43,45,49)x_4 \ge b_4\\
& x_j \ge 0 \quad \forall j
\end{aligned}$$

The location and scale parameters and specified probability levels of b_1, b_2, b_3, b_4 are given in Table 3.2.

Table 3.2: Input parameters for b_i, $i = 1,2,3,4$.

Random parameters b_i	Location parameter (δ_i)	Scale parameter (β_i)	Specified probability levels (γ_i)
b_1	800	200	0.5
b_2	600	100	0.6
b_3	150	75	0.7
b_4	86	25	0.8

Using eqs. (3.4) and (3.10), P5 can be written in the following model:

P6

$$\begin{aligned}
\text{maximize}\quad & Z_1(x) = 296.67x_1 + 287x_2 + 203x_3 + 265.67x_4\\
\text{maximize}\quad & Z_2(x) = 102.67x_1 + 121.67x_2 + 80x_3 + 85.33x_4\\
\text{subject to}\quad & 41.67x_1 + 38.67x_2 + 43.33x_3 + 22.67x_4 \le 873.3 && (3.11)\\
& 37x_1 + 21.67x_2 + 27x_3 + 32.67x_4 \le 667.17 && (3.12)
\end{aligned}$$

$$34.67x_1 + 21x_2 + 44.67x_3 + 55x_4 \geq 136.08 \tag{3.13}$$

$$42.67x_1 + 55x_2 + 64.67x_3 + 45.67x_4 \geq 74.10 \tag{3.14}$$

$$x_j \geq 0 \quad \forall j$$

Using Algorithm 3.1 of computational algorithm, P6 is converted to a linear programming problem, that is, P7.

P7

$$\begin{aligned} &\text{maximize} && \lambda \\ &\text{subject to} && \left(\frac{Z_1(x) - 618.407}{6730.487}\right) \geq \lambda \\ &&& \left(\frac{Z_2(x) - 211.122}{2661.279}\right) \geq \lambda \\ &&& \left(\frac{x_1 + x_2 - 17.363 + f}{f}\right) \geq \lambda \\ &&& \text{constraints } (3.11) - (3.14) \\ &&& x_j \geq 0 \quad \forall j \end{aligned}$$

Now, P7 is solved for $f = 0.1$ by using LINGO 15.0 iterative scheme and genetic algorithm (GA), and the optimum results are presented in Table 3.3.

Table 3.3: Optimal results for Algorithm 3.1.

Method	λ	Z_1	Z_2	x_1	x_2	x_3	x_4
LINGO	0.998	7,335.433	2,867.078	0.127	17.297	0.0	8.783
GA	0.899	7,338.571	2,848.670	0.216	16.549	0.0	9.504

Also, using Algorithm 3.2 of computational algorithm, P6 is transformed into the following model:

P8

$$\begin{aligned} &\text{maximize} && \lambda \\ &\text{subject to} && \frac{e^{-r\left(\frac{Z_1(x) - 618.407}{6730.487}\right)} - e^{-r}}{1 - e^{-r}} \geq \lambda \\ &&& \frac{e^{-r\left(\frac{Z_2(x) - 211.122}{2661.279}\right)} - e^{-r}}{1 - e^{-r}} \geq \lambda \\ &&& \frac{e^{-r\left(\frac{x_1 + x_2 - 17.363 + f}{f}\right)} - e^{-r}}{1 - e^{-r}} \geq \lambda \\ &&& \text{constraints } (3.11) - (3.14) \\ &&& x_j \geq 0 \quad \forall j \end{aligned}$$

Using LINGO 15.0 iterative scheme and GA, P8 is solved for $r = 0.5$ and $f = 0.1$, and the results are presented in Table 3.4.

Table 3.4: Optimal results for Algorithm 3.2.

Method	λ	Z_1	Z_2	x_1	x_2	x_3	x_4
LINGO	0.99999	2,717.578	1,035.560	0.00001	3.70111	3.61948	3.46520
GA	0.99956	2,718.753	1,037.992	0.00005	3.7811	3.5989	3.3989

3.5.2 Example 2

Again, let us take another example:

P9

$$\begin{aligned}
\text{maximize} \quad & Z_1(x) = (200,302,388)x_1 + (234,280,347)x_2 + (187,199,223)x_3 + (207,289,301)x_4 \\
\text{maximize} \quad & Z_2(x) = (89,107,112)x_1 + (106,127,132)x_2 + (67,80,93)x_3 + (73,79,104)x_4 \\
\text{subject to} \quad & (34,42,49)x_1 + (27,37,52)x_2 + (30,47,53)x_3 + (18,22,28)x_4 \le b_1 \\
& (32,37,42)x_1 + (18,22,25)x_2 + (25,27,29)x_3 + (30,32,36)x_4 \le b_2 \\
& (26,36,42)x_1 + (16,20,27)x_2 + (41,44,49)x_3 + (51,55,59)x_4 \ge b_3 \\
& (38,43,47)x_1 + (52,56,57)x_2 + (61,65,68)x_3 + (43,45,49)x_4 \ge b_4 \\
& x_j \ge 0 \quad \forall j
\end{aligned}$$

The location and scale parameters and specified probability levels of b_1, b_2, b_3, b_4 are given in Table 3.5.

Table 3.5: Input parameters for b_i, $i = 1, 2, 3, 4$ corresponding to Example 2.

Random parameters b_i	Location parameter (δ_i)	Scale parameter (β_i)	Specified probability levels (γ_i)
b_1	1,200	100	0.02
b_2	1,400	200	0.03
b_3	200	140	0.05
b_4	500	120	0.07

Using eqs. (3.6) and (3.10), P9 is reduced to the following model:

P10

$$\begin{aligned}
\text{maximize} \quad & Z_1(x) = 15.29x_1 + 24x_2 + 30.55x_3 + 44x_4 \\
\text{maximize} \quad & Z_2(x) = 8.7x_1 + 15.19x_2 + 6.42x_3 + 5.43x_4 \\
\text{subject to} \quad & 16.29x_1 + 34.09x_2 + 7.8x_3 + 34.21x_4 \le 1,063.59 \\
& 14x_1 + 24.4x_2 + 57.22x_3 + 54.43x_4 \le 1,149.07 \\
& 25.62x_1 + 29.3x_2 + 49.64x_3 + 54.3x_4 \ge 615.83 \\
& 4x_1 + 25.81x_2 + 67.29x_3 + 26.81x_4 \ge 814.78 \\
& x_j \ge 0 \quad \forall j
\end{aligned}$$

Using Algorithms 3.1 and 3.2 of computational algorithm, optimal results are shown in Tables 3.6 and 3.7, respectively.

Table 3.6: Optimal results for Algorithm 3.1.

Method	λ	F	Z_1	Z_2	x_1	x_2	x_3	x_4
LINGO	0.998	4.4	1,045.933	549.124	40.598	10.503	5.667	0.0
GA	0.977	4.4	1,045.983	559.914	39.985	11.897	4.88	0.0

Table 3.7: Optimal results for Algorithm 3.2.

Method	λ	F	r	Z_1	Z_2	x_1	x_2	x_3	x_4
LINGO	0.999	4.4	0.3	379.686	80.292	0.091	0.0	12.382	0.0
GA	0.995	4.4	0.3	380.318	80.329	0.074	0.0	12.412	0.0

3.5.3 Practical application (production planning problem)

India bought crude oil from different cities of the World. That crude oil cannot be directly used in our daily life. So it should be refined. Under different circumstances, the crude oil produces various by-products from different process units. India has eight oil refineries in different locations such as Haldia, Gujarat, Panipath, Paradip, Mathura, Guwahati, Barauni, Bongaigaon. Each oil refinery has different process units and are discussed as follows:

1. FOB (fuel oil block): It has two atmospheric distribution units: CDU-I and II along with the naphtha pretreating unit, catalytic reforming unit, and kero hydrodesulfurization unit.

2. DHDS (diesel-hydrodesulfurization): It produces Euro-III quality eco-friendly diesel, which reduced sulfur dioxide in vehicular emissions and as a result pollution-free environment.
3. FCCU (fluidized catalytic cracking unit): It processes heavier feedstock, including a proportion of residue, and produces value-added products and contributes to improve the distillate yield from the refinery.
4. MSQIU (motor sprit quality improvement unit): It produces EURO-III petrol.
5. OHCU (once-through hydrocracker): It comprises hydrogen generation unit, sulfur recovery units, hydrogen storage facility, air compressor, heat recovery steam generation, effluent treatment plant, and feed preparation unit.
6. LOB (lube oil block): It consists of vacuum distillation unit, propane deasphalting unit, lube hydrofinishing unit, vis-breaking unit and catalytic ISO dewaxing unit.
7. DHDT (diesel hydrotreating unit): It produces EURO-IV-grade MS and HSD (High Speed Diesel).
8. GT (gas turbines): It generates electricity by using gas turbine.

All the process units are not available in all refineries. Due to fluctuation of global market and plant equipments, the capacity of the process units in each refinery is not fixed. So considering these capacities are fuzzy numbers (triangular or trapezoidal). We mainly discuss here six types of oil refineries and their capacities that are supplied in Tables 3.8 and 3.9, respectively.

Table 3.8: Input data for FOB, DHDS, FCCU, and MSQIU units.

Refineries	Process units			
	FOB	DHDS	FCCU	MSQIU
Haldia	(5,5.5,7)	(6,6.3,6.5)	(10,12,12.3)	(11,11.2,11.4)
Gujarat	(1.1,1.4,1.5)	(2.1,2.3,2.6)	(2.4,2.41,2.43)	(2.13,2.16,2.17)
Panipath	(1.9,2.3,2.7)	–	(1.2,1.3,1.7)	–
Paradip	(0.47,0.49,1)	(0.57,1,1.1)	–	(0.39,0.45,0.5)
Mathura	(5.6,5.86,5.89)	(6.2,6.5,7.2)	(6.15,6.19,6.34)	(5.1,5.15,5.24)
Guwahati	(2.12,2.15,2.17)	–	(2.18,2.19,2.21)	(1.89,1.92,1.95)

Let us consider that x_1, x_2, x_3, x_4, x_5, x_6, x_7, and x_8 be the decision variables (quantity) corresponding to the process units FOB, DHDS, FCCU, MSQIU, OHCU, LOB, DHDT, and GT, respectively. So the problem gives a TLPP where the top-level DM maximizes the profit and the lower one also maximizes the products. The cost

Table 3.9: Input data for OHCU, LOB, DHDT, and GT units.

Refineries	Process units			
	OHCU	LOB	DHDT	GT
Haldia	(13,13.2,13.5)	(12,12.6,14)	(7.2,7.5,7.9)	(5.5,5.9,6.3)
Gujarat	(1.8,2.1,2.3)	(1.12,1.14,1.17)	(1.43,1.45,1.48)	(2.6,2.9,3.1)
Panipath	(2.1,3.3,3.4)	(3.21,3.42,3.46)	–	(3.42,3.45,4)
Paradip	–	(0.31,0.38,0.47)	(0.67,0.69,0.72)	–
Mathura	(4.3,4.4,4.49)	–	(4.18,4.24,4.38)	(4.67,4.75,4.86)
Guwahati	–	(2.15,2.19,2.27)	(1.73,1.79,1.83)	–

coefficients of both level objective functions and location, shape, and specified probability levels of oil refineries are shown in Tables 3.10 and 3.11, respectively. Using Tables 3.8–3.11, model P3 is transformed into the following model:

Table 3.10: Input data for cost coefficients.

$c_{11} = (800, 820, 850)$	$c_{12} = (380, 420, 435)$	$c_{13} = (305, 310, 325)$
$c_{14} = (210, 225, 240)$	$c_{15} = (180, 197, 213)$	$c_{16} = (170, 175, 180)$
$c_{17} = (150, 156, 159)$	$c_{18} = (140, 148, 153)$	$c_{21} = (80, 85, 89)$
$c_{22} = (52, 53, 55)$	$c_{23} = (46, 48, 49)$	$c_{24} = (32, 35, 38)$
$c_{25} = (27, 27.6, 28)$	$c_{26} = (17, 17.6, 17.9)$	$c_{27} = (12, 12.9, 13)$
$c_{28} = (8, 8.7, 8.9)$		

Table 3.11: Input parameters for location, shape, and probability levels.

Refineries	Location parameter (δ_i)	Scale parameter (β_i)	Specified probability levels (γ_i)
Haldia	200	20	0.05
Gujarat	64	10	0.06
Panipath	19	9	0.07
Paradip	6	3	0.07
Mathura	5	3	0.08
Guwahati	7	2	0.09

P11

$$
\begin{aligned}
\text{maximize} \quad & Z_1(x) = 823.33x_1 + 411.67x_2 + 313.33x_3 + 225x_4 + 196.67x_5 + 175x_6 + 155x_7 + 147x_8 \\
\text{maximize} \quad & Z_2(x) = 84.67x_1 + 53.33x_2 + 47.67x_3 + 35x_4 + 27.53x_5 + 17.5x_6 + 12.63x_7 + 8.53x_8 \\
\text{subject to} \quad & 5.83x_1 + 6.27x_2 + 11.43x_3 + 11.2x_4 + 13.23x_5 + 12.87x_6 + 7.53x_7 + 5.9x_8 \le 178.06 \\
& 1.33x_1 + 2.33x_2 + 2.41x_3 + 2.15x_4 + 2.07x_5 + 1.14x_6 + 1.45x_7 + 2.87x_8 \le 53.66 \\
& 2.3x_1 + 1.4x_3 + 2.93x_5 + 3.36x_6 + 3.62x_8 \le 10.20 \\
& 0.65x_1 + 0.89x_2 + 0.45x_4 + 0.39x_6 + 0.69x_7 \ge 13.87 \\
& 5.78x_1 + 6.63x_2 + 6.23x_3 + 5.14x_4 + 4.4x_5 + 4.27x_7 + 4.76x_8 \ge 12.45 \\
& 2.15x_1 + 2.19x_3 + 1.92x_4 + 2.2x_6 + 1.78x_7 \ge 11.72 \\
& x_j \ge 0 \quad \forall j
\end{aligned}
$$

Using Algorithms 3.1 and 3.2 of computational algorithm, optimal solutions are obtained from P11 by LINGO 15.0 iterative scheme and GA and put into Tables 3.12 and 3.13, respectively.

Table 3.12: Optimal results for Algorithm 3.1.

Method	Z_1	Z_2	x_1	x_2	x_3	x_4	x_5	x_6	x_7	x_8
LINGO	11,936.45	1,448.545	4.435	19.574	0.0	0.64	0.0	0.0	0.537	0.0
GA	12,141.87	1,448.597	5.568	17.774	0.0	0.58	0.0	0.0	0.71	0.0

Table 3.13: Optimal results for Algorithm 3.2.

Method	Z_1	Z_2	x_1	x_2	x_3	x_4	x_5	x_6	x_7	x_8
LINGO	3,447.215	304.995	0.0	0.0	0.0	1.542	0.0	0.322	18.914	0.764
GA	3,447.324	306.866	0.0	0.0	0.0	1.627	0.0	0.413	18.728	0.722

3.6 Discussion

In our chapter, we solved a stochastic TLPP for a noncooperative game under fuzzy environment using linear and nonlinear membership functions. For Example 1, from Tables 3.3 and 3.4, it has been observed that the optimal solutions have been derived using LINGO 15.0 iterative scheme and GA, respectively. In case of linear membership function under GA, we have the best solution for $\lambda = 0.899$ and $d = 0.1$ corresponding to both level DMs $Z_1 = 7338.571$, $Z_2 = 2848.670$. However, for the nonlinear membership function under GA, the results are $Z_1 = 2718.753$, $Z_2 = 1037.992$ with $\lambda = 0.99998$, $t = 0.5$ and $d = 0.1$, respectively. For Example 2, from Tables 3.6 and 3.7, it is concluded that the optimal solutions are reached only for the linear membership function under GA $Z_1 = 1045.983$, $Z_2 = 559.914$ with $\lambda = 0.977$ and $d = 4.4$. Again, for the production planning problem, from Tables 3.12 and 3.13, we have observed that the optimal results have been obtained for both levels $Z_1 = 12141.87$, $Z_2 = 1448.597$, $d = 4.4$, $\lambda = 0.999$ corresponding to the linear membership function, whereas $Z_1 = 3447.324$, $Z_2 = 306.866$, $d = 1.1$, $t = 0.2$, and $\lambda = 0.9567$ corresponding to the nonlinear membership function. So the best solution occurred corresponds to the linear membership function under GA. Finally, we have concluded that the three numerical experiments that attained the best solution correspond to only consider the linear membership function under GA.

3.7 Concluding remarks and future research

In this chapter we considered a TLPP for a noncooperative game where the cost parameters and the constraint parameters except on the right-hand side are either triangular or trapezoidal fuzzy numbers, whereas the constraints on the right side follow an extreme value distribution. Using Mellin transformation, fuzzy numbers are transformed into crisp value. Thereafter, the constraints in probabilistic nature have been converted into a crisp form by using the stochastic programming procedure. So it becomes a nonlinear programming problem. Then based on the computational algorithm, reduced problem is solved by LINGO 15.0 iterative scheme and GA, respectively and then compared the results.

The following aspects are treated as new features of our chapter.

(a) For the first time, we have proposed a probabilistic TLPP in Stackelberg game under a fuzzy environment using linear and nonlinear membership functions.
(b) Based on Mellin transformation, all types of fuzzy numbers (triangular or trapezoidal) are converted into crisp values.
(c) Probabilistic constraints follow an extreme value distribution to be converted into a deterministic form using stochastic programming.
(d) Comparative study about the results by LINGO 15.0 iterative scheme and GA correspond to nonlinear and linear memberships.

In future studies, this method can be considered in another type of nonlinear memberships such as hyperbolic, parabolic, and logarithm. This research work can be extended for multiobjective TLPP in noncooperative game, including managerial decision-making problem and transportation problem.

References

[1] Anandalingam G. & Apprey V. (1991) Multi-level programming and conflict resolution, European Journal of Operational Research, 51(2), 233–247.

[2] Maiti S.K. & Roy S.K. (2018) Bi-level programming for Stackelberg game with intuitionistic fuzzy number: a ranking approach, Journal of the Operations Re- search Society of China. https://doi.org/10.1007/ s40305-018-0234-2.

[3] Roy S.K. (2006) Fuzzy programming techniques for Stackelberg game, Tamsui Oxford Journal of Management Sciences, 22(3), 43–56.

[4] Roy S.K. & Maiti S.K. (2017) Stochastic bi level programming with multi-choice for Stackelberg game via fuzzy programming, International Journal of Operational Research, 29 (4), 508–530.

[5] Roy S.K., Maity G., & Weber G.W. (2017) Multi-objective two-stage grey transportation problem using utility function with goals, Central European Journal of Operations Research, 25(2), 417–439.

[6] Roy S.K. (2014) Multi-choice stochastic transportation problem involving Weibull distribution, International Journal of Operational Research, 21(1), 38–58.

[7] Anandalingam G. & White D.J. (1990) A solution method for the linear static Stackelberg problem using penalty functions, IEEE Transactions on Automatic Control, 35(10), 1170–1173.

[8] Hamidi F. & Mishmast N. (2013) Bi-level linear programming with fuzzy parameters, Iranian Journal of Fuzzy Systems, 10(4), 83–99.

[9] Sakawa M., Katagiri H., & Matsui T. (2012) Fuzzy random bi-level linear programming through expectation optimization using possibility necessity, International Journal of Machine Learning and Cybernetics, 3(3), 183–192.

[10] Singh V.P. & Chakraborty D. (2017) Solving bi-level programming problem with fuzzy random variable coefficients, Journal of Intelligent and Fuzzy Systems, 32(1), 521–528.

[11] Yeh K., Whittaker C., Realffa M.J., & Lee J.H. (2015) Two stage stochastic bi-level programming model of a pre-established timber lands supply chain with biorefinery investment interests, Computers and Chemical Engineering, 73(2), 141–153.

[12] Youness E.A., Emam O.E., & Hafez M.S. (2014) Fuzzy bi- level multi- objective fractional integer programming, Applied Mathematics and Information Sciences, 8(6), 2857–2863.

[13] Zheng Y., Wana Z., & Wang G. (2011) A fuzzy interactive method for a class of bi-level multi-objective programming problem, Expert Systems with Applications, 38(8), 10384–10388.

[14] Grzegorzewski P. (2002) Nearest interval approximation of a fuzzy number, Fuzzy Sets and Systems, 130(3), 321–330.

[15] Debnath L. (1965) Integral transforms and their applications, CRC Press, New York.

[16] Zimmermann H.J. (1978) Fuzzy programming and linear programming with several objective functions, Fuzzy Sets and Systems, 1(1), 45–55.

[17] Roy S.K. (2007) Fuzzy programming approach to two person multi-criteria bi-matrix games, The Journal of Fuzzy Mathematics, 15(1), 141–153.

[18] Zadeh L.A. (1965) Fuzzy sets, Information and Control, 8(3), 338–353.

[19] Ren A.H. & Wang Y.P. (2014) Optimistic Stackelberg solutions to bi-level linear programming with fuzzy random variable coefficients, Knowledge Based Systems. https://doi.org/10.1016/j.knosys.2014.05.010.

[20] Maiti S.K. & Roy S.K. (2016) Multi-choice stochastic bi-level programming problem in co-operative nature via fuzzy programming approach, Journal of Industrial Engineering International, 12(3), 287–298.

[21] Lai Y.J. (1996) Hierarchical optimization: A satisfactory solution, Fuzzy Sets and Systems, 77 (3), 321–335.

[22] Mahapatra D.R., Roy S.K., & Biswal M.P. (2013) Multi-choice stochastic transportation problem involving extreme value distribution, Applied Mathematical Modelling, 37(4), 2230–2240.
[23] Kiyota S. & Aiyishi E. (1981) A computational method for Stackelberg and min max problem use of a penalty method, IEEE Transactions on Automatic Control, 26(2), 460–466.
[24] Sakawa M. & Matsui T. (2016) Bi-level linear programming with fuzzy random variables through absolute deviation minimization, International Journal of Operational Research, 25 (1), 1–27.
[25] Sakawa M. & Yano H. (1989) An interactive fuzzy satisfying method of multi- objective nonlinear programming problems with fuzzy parameters, Fuzzy Sets and Systems, 30(3), 221–238.

D. K. Sharma, Rajnee Tripathi

4 Intuitionistic fuzzy trigonometric distance and similarity measure and their properties

Abstract: In this chapter, we introduce the concept of intuitionistic fuzzy sets (IFS). Similarity and distance measures between IFS are explained and also extended these measures to IFSs. Some new trigonometric similarity and distance measures of IFSs are studied, and it is shown that intuitionistic fuzzy distance measures satisfy the required identities of intuitionistic fuzzy similarity measures. Also, basic properties of trigonometric similarity and distance measures of IFS (IF_s) are described.

Keywords: Fuzzy set (F_S), intuitionistic fuzzy set (IF_S), distance measures (DM), similarity measures (SM), intuitionistic fuzzy distance measures (IFDM), intuitionistic fuzzy similarity measures (IFSM)

Subject classification codes (2010): 94D05, 94A15, 94A17.

4.1 Introduction

The concept of fuzzy set was proposed by Zadeh [1]. Fuzzy sets conjecture has to transact business with uncertainty and dreaminess anywhere the notch of belongingness of any part to a backdrop which is represented by a attachment function. However, in the frank mean condition, it is not constantly doable that the scale of nonmembership meeting of a facet to the fuzzy set is barely peer to 1 minus the level of membership. It implies that there is some kind of degree of ambiguity and has been called hesitation degree.

The classical concept of fuzzy sets [1] is a strong tool to deal with uncertainty, vagueness, and imprecise information. There are various generalizations of higher order fuzzy sets for different specialized purposes. Intuitionistic fuzzy sets (IFS) introduced by Atanassov [2] are quite useful and applicable, and fuzzy sets are described by two functions: a membership function and a nonmembership function. Atanassov [2] introduced the concept of IFS as simplification of fuzzy sets, which is competent of capturing the order that includes several grades of timid and has been practical in countless return of research.

D. K. Sharma, Rajnee Tripathi, Jaypee University of Engineering and Technology, Madhya Pradesh, India

https://doi.org/10.1515/9783110628616-004

Thus, another degree of freedom has been introduced to describe IFSs. Such a generalization of fuzzy sets equips us with an additional possibility to represent incomplete knowledge, which leads to describe many real problems in a more adequate way. Further, Attanssov and Gargov [3] also introduced the concept of interval-valued IFS and studied their properties.

Atanassov [4] provided an example, where fuzzy sets are IFS; however, converse is not automatically rightful. Burillo and Bustince [5] suggested an algorithm for IFS from fuzzy sets with the help of Atanassov's [6] operator. Each of the issues that manage fuzzy set hypothesis can also manage IFS hypothesis; however, there are circumstances where fuzzy set hypothesis cannot be appropriate to make a reasonable investigation. The membership values cannot be always determined, due to inefficiency of available information, besides the presence of vagueness in the information. Similarly, while determining the nonmembership, the same situation arises. A part of such estimation remains of indeterministic nature. This indeterministic part has not been considered in the case of fuzzy set theory, so we have IFS. For instance, suppose that a human being is to express his/her opinion about the beauty of a flower. In classical theory, he/she has only two choices 0 or 1, that is, a flower is either beautiful or ugly. In fuzzy sets, he/she has the liberty to express his/her view that the flower is beautiful with a grade of membership 0.8, which determines that the grade of nonmembership is 0.2, but in real life, a flower may have the beauty with a grade of membership 0.8 but not ugly at all, that is, its grade of nonmembership has the value 0. Therefore, one needs a most appropriate approach, which would help to deal with such a problem, whose situation depends on human perception. Therefore, IFS is a very useful mathematical tool in many situations having imprecise, vague, and inexact data.

It is noted that IFSs can describe the fuzzy characters of the things more comprehensively and thus are powerful and effective tools in dealing with fuzzy information.

An intuitionistic fuzzy conventional is characterized by two functions. Former is the point of link and subsequently is the level of nonmembership function. Their addition is each time fewer than one and the same to one.

Definition 4.1 (Fuzzy Set).
Fuzzy set E in a universal set Y is defined as

$$E = \{< y, \alpha_E(y) >: y \in E\} \tag{4.1}$$

where $\alpha_E(y_i)$ is a membership function defined as follows:

$$\alpha_E(y) = \begin{cases} 0, & y \text{ does not belong to } E \text{ and there is no ambiguity} \\ 1, & y \text{ belongs to } E \text{ and there is no ambiguity} \\ 0.5, & \text{there is maximum ambiguity whether } y \text{ belongs to } E \text{ or not} \end{cases} \tag{4.2}$$

In fact $\alpha_E(y)$ associates with each y, a grade of membership, to the set E.

Definition 4.2 (Intuitionistic Fuzzy Set).
An IFS E in a universal set Y is defined as

$$E = \{< y, \alpha_E(y), \beta_E(y) >: y \in E\} \tag{4.3}$$

where the functions

$$E = \{< y, \alpha_E(y), \beta_E(y) >: y \in E\} \tag{4.4}$$

are the degree of membership to the element and the degree of nonmembership to the element, respectively, with the condition

$$0 \leq \alpha_E(y) + \beta_E(y) \leq 1 \tag{4.5}$$

the intuitionistic index of an element $y \in Y$ is defined as follows:

$$\pi_E(y) = 1 - \alpha_E(y) - \beta_E(y) \tag{4.6}$$

$\pi_A(x)$ is called a degree of membership of an element in E and it is evident that $0 \leq \pi_E(y) \leq 1$.

Note that for an IFS E, if $\alpha_E(y) = 0$ then $\alpha_E(y) + \beta_E(y)$ and if $\alpha_E(y) = 1$ then $\pi_E(y) = 0$.

We present the following basic operations on IFS, which is necessary for our discussion:

Definition 4.3 (Basic Operations of Intuitionistic Fuzzy Set). If E and F are two IFS of the universal set Y, then

(i) iff $E \subset F$ then $\alpha_E(y) \leq \alpha_F(y)$ and $\beta_E(y) \geq \beta_F(y)$ $\forall y \in Y$

(ii) iff $E = F$ then $\alpha_E(y) = \alpha_F(y)$ and $\beta_E(y) = \beta_F(y)$

(iii) $E^C = \{< y, \beta_E(y), \alpha_E(y) >: y \in Y\}$, where E^C is complement of E

(iv) $E \cap F = \{< y, \min(\alpha_E(y), \alpha_F(y)), \max(\alpha_E(y), \alpha_F(y)) >: y \in Y\}$

(v) $E \cup F = \{< y, \max(\alpha_E(y), \alpha_F(y)), \min(\alpha_E(y), \alpha_E(y)) >: y \in Y\}$

Burillo and Bustince [5] have also given some results on IFS. In the subsequent discussion, the following notations are used:

$R = [0, \infty)$, $Y = \{y_1, y_2, \ldots, y_n\}$ is the universal set; $F_s(Y)$ is the class of all fuzzy sets of Y; IFS(Y) is the class of all IFSs of Y.

Liu [7] given new measures without saying definitions of aloofness actions and similarity events of fuzzy sets and examined the connections between them. These axioms are spontaneous and possess laboring in the fuzzy literature.

Definition 4.4 (Fuzzy Distance Measure). Let D be a real-valued function defined as $D: F_S \times F_S \to R^+$. A distance measure D is called fuzzy distance measure if it satisfies the following conditions:

(d1) $D(E,F) = D(F,E) \ \ \forall E,F \in F_s$

(d2) $D(E,E) = 0 \ \ \forall E \in F_s$

(d3) $D(d,d^C) = \max_{E,F \in F_S} D(E,F)$, if d is the crisp set

(d4) $\forall E,F,G \in F_S$, if $E \subset F \subset G$ then $D(E,F) \le D(E,G)$ and $D(E,F) \le D(F,G)$

Definition 4.5 (Fuzzy Similarity Measure). Suppose that S be a real-valued function $T: F_S \times F_S \to R^+$. A similarity measure T is called fuzzy similarity measure if it satisfies the following conditions:

(S1) $T(E,F) = T(F,E) \ \ \forall E,F \in F_s$

(S2) $T(d,d^C) = 0$, if d is the crisp set

(S3) $T(\mathrm{K},K) = \max_{E,E \in F_S} D(E,F) \ \forall K F$

(S4) $\forall E,F,G \in F_S$, if $E \subset F \subset G$ then $T(E,F) \ge T(E,G)$ and $T(E,F) \ge T(F,G)$

If we standardize d and T in Definitions 4.3 and 4.4, we get $0 \le D(E,F) \le 1$ and $0 \le T(E,F) \le 1$ for E, $F \in F_S$. The relation between standardized d and S is $D = 1 - T$. Hence, from this relation, it can be seen that the distance and similarity measures are dual concepts, and we can give the distance and similarity measure axiomatic definitions for IFS.

Definition 4.6 (Intuitionistic Fuzzy Distance Measure). Suppose D is a real-valued function $D: \mathrm{IFS} \times \mathrm{IFS} \to R^+$. Then distance measure D is called intuitionistic fuzzy distance measure if it satisfies the following conditions:

(IS1) $T(E,F) = T(F,E) \ \ \forall E,F \in \mathrm{IF}_s$

(IS2) $T(d,d^C) = 0$, if is the crisp set

(IS3) $T(K,K) = \max_{E,F \in \mathrm{IF}_S} D(E,F) \ \ \forall K \in IF_s$

(IS4) $\forall E,F,G \in \mathrm{IF}_S$, if $E \subset F \subset G$ then $T(E,F) \ge T(E,G)$ and $T(E,F) \ge T(F,G)$

Definition 4.7 (Intuitionistic Fuzzy Similarity Measure). Let T be a real-valued function $T: \mathrm{IF}_S \times \mathrm{IF}_S \to R^+$. Then similarity measure T is called fuzzy similarity measure if it satisfies the following conditions:

(IS1) $T(E,F)=T(F,E)\quad \forall E,F\in \text{IFS}$

(IS2) $T(d,d^C)=0$, if d is the crisp set

(IS3) $T(K,K)=\max_{E,F\in \text{IF}_S} D(E,F)\quad \forall K\in IF_S$

(IS4) $\forall E,F,G\in \text{IF}_S$, if $E\subset F\subset G$ then $T(E,F)\geq T(E,G)$ and $T(E,F)\geq T(F,G)$

4.2 Sine trigonometric intuitionistic fuzzy distance measures

Definition 4.8. Let E and F be two IFS on universal set Y as $Y=\{y_1,y_2,\ldots,y_n\}$, whose membership values $\alpha_E(y_i),\ i=1,2,\ldots,n$ and $\alpha_F(y_i),\ i=1,2,\ldots,n$, respectively, are the measures of sine intuitionistic fuzzy distance $D_{\text{IF}_S}(E,F)$ between $D_{\text{IF}_S}(E,F)$ fuzzy sets E and F, as

$$D_{\text{IF}_{S1}}(E,F)=\frac{1}{2n}\sum_{i=1}^{n}\left(\sin\left(\frac{|\alpha_E(y_i)-\alpha_F(y_i)|}{2}\right)\pi+\sin\left(\frac{|\beta_E(y_i)-\beta_F(y_i)|}{2}\right)\pi\right) \tag{4.7}$$

$$D_{\text{IF}_{S2}}(E,F)=\frac{1}{2n}\sum_{i=1}^{n}\left[\sin\left(\frac{|\sqrt{\alpha_E(y_i)}-\sqrt{\alpha_F(y_i)}|}{2}\right)\pi+\sin\left(\frac{|\sqrt{\beta_E(y_i)}-\sqrt{\beta_F(y_i)}|}{2}\right)\pi\right] \tag{4.8}$$

Theorem 4.1. *Identities* (4.7) *and* (4.8) *prove all the conditions of intuitionistic fuzzy distance measures.*

Proof. Given that

$$D_{\text{IF}_{S1}}(E,F)=\frac{1}{2n}\sum_{i=1}^{n}\left[\sin\left(\frac{|\alpha_E(y_i)-\alpha_F(y_i)|}{2}\right)\pi+\sin\left(\frac{|\beta_E(y_i)-\beta_F(y_i)|}{2}\right)\pi\right]$$

Here, we prove all four conditions of intuitionistic fuzzy distance measures that are mentioned in Definition 4.6:

(Id1)
$$\begin{aligned}D_{\text{IF}_{S1}}(E,F)&=\frac{1}{2n}\sum_{i=1}^{n}\left[\sin\left(\frac{|\alpha_E(y_i)-\alpha_F(y_i)|}{2}\right)\pi+\sin\left(\frac{|\beta_E(y_i)-\beta_E(y_i)|}{2}\right)\pi\right]\\&=\frac{1}{2n}\sum_{i=1}^{n}\left[\sin\left(\frac{|\alpha_F(y_i)-\alpha_E(y_i)|}{2}\right)\pi+\sin\left(\frac{|\beta_F(y_i)-\beta_E(y_i)|}{2}\right)\pi\right]\\&=D_{\text{IF}_{S1}}(F,E)\end{aligned}$$

(Id2) $D_{\mathrm{IF_{S1}}}(E,E)=\frac{1}{2n}\sum_{i=1}^{n}\left[\sin\left(\frac{|\alpha_E(y_i)\ -\alpha_E(y_i)|}{2}\right)\pi+\sin\left(\frac{|\beta_E(y_i)\ -\beta_E(y_i)\,|}{2}\right)\pi\right]$

$=0\quad\forall E,F\in \mathrm{IF_S}$

(Id3) Let

$E=d$ and $F=d^C$, i.e., $\alpha_E(y_i)=\alpha_d(y_i)$, $\alpha_F(y_i)=\alpha_{d^C}(y_i)$ and $\beta_E(y_i)=\beta_d(y_i)$, $\beta_B(y_i)=\beta_{d^C}(y_i)$

Then we get

$$D_{\mathrm{IF_{S1}}}(d,d^C)=1\quad\forall E,F\in \mathrm{IF_S}$$

that is

$$D_{\mathrm{IF_{S1}}}(d,d^C)=\max_{E,F\in \mathrm{IF_S}} D_{\mathrm{IF_{S1}}}(E,F)\quad \forall E,F\in \mathrm{IF_S}.$$

(Id4) For all $E,F,G\in \mathrm{IF_S}(Y)$, if $E\subset F\subset G$, then $\alpha_E(y_i)\le\alpha_F(y_i)\le\alpha_G(y_i)$ or $\beta_E(y_i)\ge\beta_F(y_i)\ge\beta_G(y_i)$

$$\left(\frac{|\alpha_E(y_i)-\alpha_F(y_i)|}{2}\right)\pi+\left(\frac{|\beta_E(y_i)-\beta_F(y_i)\,|}{2}\right)\pi\le\left(\frac{|\alpha_E(y_i)-\alpha_G(y_i)|}{2}\right)\pi+\left(\frac{|\beta_E(y_i)-\beta_G(y_i)\,|}{2}\right)\pi$$

and the nature of sine function, we get

$$\sin\left(\frac{|\alpha_E(y_i)-\alpha_F(y_i)|}{2}\right)\pi+\sin\left(\frac{|\beta_E(y_i)-\beta_F(y_i)\,|}{2}\right)\pi\le\sin\left(\frac{|\alpha_E(y_i)-\alpha_G(y_i)|}{2}\right)\pi+\sin\left(\frac{|\beta_E(y_i)-\beta_G(y_i)\,|}{2}\right)\pi$$

$\Rightarrow D_{\mathrm{IF_{S1}}}(E,F)\le D_{\mathrm{IF_{S1}}}(E,G)$

Similarly,

$$\left(\frac{|\alpha_E(y_i)-\alpha_F(y_i)|}{2}\right)\pi+\left(\frac{|\beta_E(y_i)-\beta_F(y_i)\,|}{2}\right)\pi\le\left(\frac{|\alpha_F(y_i)-\alpha_G(y_i)|}{2}\right)\pi+\left(\frac{|\beta_F(y_i)-\beta_G(y_i)\,|}{2}\right)\pi$$

$$\sin\left(\frac{|\alpha_E(y_i)\ -\alpha_F(y_i)|}{2}\right)\pi+\sin\left(\frac{|\beta_E(y_i)\ -\beta_F(y_i)\,|}{2}\right)\pi\le\sin\left(\frac{|\alpha_E(y_i)\ -\alpha_G(y_i)|}{2}\right)\pi$$
$$+\sin\left(\frac{|\beta_E(y_i)\ -\beta_G(y_i)\,|}{2}\right)\pi$$

$$\Rightarrow D_{\mathrm{IF_S}}(E,F)\le D_{\mathrm{IF_S}}(F,G)$$

Now we will prove all the required conditions using identity given in eq. (4.8):

$$\text{(Id1) } D_{\mathrm{IF_{S2}}}(E,F)=\frac{1}{2n}\sum_{i=1}^{n}\left[\left(\frac{|\sqrt{\alpha_E(y_i)}-\sqrt{\alpha_F(y_i)}|}{2}\right)\pi+\sin\left(\frac{|\sqrt{\beta_E(y_i)}-\sqrt{\beta_F(x_i)}\,|}{2}\right)\pi\right]$$
$$=\frac{1}{2n}\sum_{i=1}^{n}\left[\sin\left(\frac{|\sqrt{\alpha_F(y_i)}-\sqrt{\alpha_E(y_i)}|}{2}\right)\pi+\sin\left(\frac{|\sqrt{\beta_F(y_i)}-\sqrt{\beta_E(y_i)}\,|}{2}\right)\pi\right]$$
$$=D_{\mathrm{IF_{S2}}}(F,E)$$

$$\text{(Id2) } D_{\mathrm{IF_{S2}}}(E,E)=\frac{1}{2n}\sum_{i=1}^{n}\left[\sin\left(\frac{|\sqrt{\alpha_E(y_i)}-\sqrt{\alpha_E(y_i)}|}{2}\right)\pi+\sin\left(\frac{|\sqrt{\beta_E(y_i)}-\sqrt{\beta_E(y_i)}\,|}{2}\right)\pi\right]$$
$$=0\ \ \forall E\in \mathrm{IF_S}$$

(Id3) Let

$E=d$ and $F=d^C$, i.e., $\alpha_E(y_i)=\alpha_d(y_i)$, $\alpha_F(y_i)=\alpha_{d^C}(y_i)$ and $\beta_E(y_i)=\beta_d(y_i)$, $\beta_F(y_i)=\beta_{d^C}(y_i)$

Then we get,

$$D_{\mathrm{IF_{S2}}}(d,d^C)=1\quad\forall E,F\in\mathrm{IF_S}$$

that is, $D_{\mathrm{IF_{S2}}}(d,d^C)=\max_{E,F\in\mathrm{IF_S}}D_{\mathrm{IF_{S2}}}(E,F)\ \ \forall E,F\in\mathrm{IF_S}$.

(Id4) For all $E,F,G\in\mathrm{IF_S}(Y)$, if $E\subset F\subset G$, then $\alpha_E(y_i)\le\alpha_F(y_i)\le\alpha_G(y_i)$ or $\beta_E(y_i)\ge\beta_F(y_i)\ge\beta_G(y_i)$

$$\left(\frac{|\sqrt{\alpha_E(y_i)}\ -\sqrt{\alpha_F(y_i)}|}{2}\right)\pi+\left(\frac{|\sqrt{\beta_E(y_i)}\ -\sqrt{\beta_F(y_i)}\,|}{2}\right)\pi\le\left(\frac{|\sqrt{\alpha_E(y_i)}\ -\sqrt{\alpha_G(y_i)}|}{2}\right)\pi$$
$$+\left(\frac{|\sqrt{\beta_E(y_i)}\ -\sqrt{\beta_G(y_i)}\,|}{2}\right)\pi$$

and the nature of sine function must be defined as

$$\sin\left(\frac{|\sqrt{\alpha_E(y_i)} - \sqrt{\alpha_F(y_i)}|}{2}\right)\pi + \sin\left(\frac{|\sqrt{\beta_E(y_i)} - \sqrt{\beta_F(y_i)}|}{2}\right)\pi$$

$$\leq \sin\left(\frac{|\sqrt{\alpha_E(y_i)} - \sqrt{\alpha_G(y_i)}|}{2}\right)\pi + \sin\left(\frac{|\sqrt{\beta_E(y_i)} - \sqrt{\beta_G(y_i)}|}{2}\right)\pi.$$

$\Rightarrow D_{\mathrm{IF}_{S2}}(E,F) \leq D_{\mathrm{IF}_{S2}}(E,G)$

Similarly,

$$\left(\frac{|\sqrt{\alpha_E(y_i)} - \sqrt{\alpha_F(y_i)}|}{2}\right)\pi + \left(\frac{|\sqrt{\beta_E(y_i)} - \sqrt{\beta_F(y_i)}|}{2}\right)\pi \leq \left(\frac{|\sqrt{\alpha_F(y_i)} - \sqrt{\alpha_G(y_i)}|}{2}\right)\pi$$

$$+ \left(\frac{|\sqrt{\beta_F(y_i)} - \sqrt{\beta_G(y_i)}|}{2}\right)\pi$$

$$\Rightarrow \sin\left(\frac{|\sqrt{\alpha_E(y_i)} - \sqrt{\alpha_F(y_i)}|}{2}\right)\pi + \sin\left(\frac{|\sqrt{\beta_E(y_i)} - \sqrt{\beta_F(y_i)}|}{2}\right)\pi$$

$$\leq \sin\left(\frac{|\sqrt{\alpha_F(y_i)} - \sqrt{\alpha_G(y_i)}|}{2}\right)\pi + \sin\left(\frac{|\sqrt{\beta_F(y_i)} - \sqrt{\beta_G(y_i)}|}{2}\right)\pi.$$

$\Rightarrow D_{\mathrm{IF}_{S2}}(E,F) \leq D_{\mathrm{IF}_{S2}}(F,G)$

4.3 Properties of sine trigonometric intuitionistic fuzzy distance measures

Let D be a distance measure on IFS and for $E, F \in \mathrm{IF}_S$. If they satisfy that for all $y_i \in Y$, either $E(y_i) \subseteq F(y_i)$ *or* $E(y_i) \supseteq F(y_i)$, then

$$\text{(i) } D_{\mathrm{IF}_{S1}}(E \cup F, E \cap F) = D_{\mathrm{IF}_{S1}}(E,F) \tag{4.9}$$

and

$$D_{\mathrm{IF}_{S2}}(E \cup F, E \cap F) = D_{\mathrm{IF}_{S2}}(E,F) \quad \forall E,F,G \in \mathrm{IF}_S \tag{4.10}$$

$$\text{(ii) } D_{\mathrm{IF}_{S1}}(E \cup F, G) \leq D_{\mathrm{IF}_{S1}}(E,G) + D_{\mathrm{IF}_{S1}}(F,G) \tag{4.11}$$

and

$$D_{\mathrm{IF}_{S2}}(E \cup F, G) \leq D_{\mathrm{IF}_{S2}}(E,G) + D_{\mathrm{IF}_{S2}}(F,G) \quad \forall E,F,G \in \mathrm{IF}_S \tag{4.12}$$

$$\text{(iii) } D_{\mathrm{IF}_{S1}}(E\cap F,G)\leq D_{\mathrm{IF}_{S1}}(E,G)+D_{\mathrm{IF}_{S1}}(F,G) \tag{4.13}$$

$$\text{and } D_{\mathrm{IF}_{S2}}(E\cap F,G)\leq D_{\mathrm{IF}_{S2}}(E,G)+D_{\mathrm{IF2}}(F,G)\quad \forall E,F,G\in \mathrm{IF_S} \tag{4.14}$$

then the above equations satisfy the properties.

Proof.

(i) Using Definition 4.6, we get

$$D_{\mathrm{IF}_{S1}}(E\cup F,E\cap F)=\frac{1}{2n}\sum_{i=1}^{n}\left[\sin\left(\frac{|\alpha_{E\cup F}(y_i)-\alpha_{E\cap F}(y_i)|}{2}\right)\pi\right.$$
$$\left.+\sin\left(\frac{|\beta_{E\cup F}(y_i)-\beta_{E\cap F}(y_i)|}{2}\right)\pi\right]$$
$$=\frac{1}{2n}\left[\sum_{y_i\in Y_1}\left[\sin\left(\frac{|\alpha_F(y_i)-\alpha_E(y_i)|}{2}\right)\pi+\sin\left(\frac{|\beta_F(y_i)-\beta_E(y_i)|}{2}\right)\pi\right]\right.$$
$$\left.+\sum_{y_i\in Y_2}\left[\sin\left(\frac{|\alpha_E(y_i)-\alpha_F(y_i)|}{2}\right)\pi+\sin\left(\frac{|\beta_E(y_i)-\beta_F(y_i)|}{2}\right)\pi\right]\right]$$
$$=\frac{1}{2n}\sum_{i=1}^{n}\left[\sin\left(\frac{|\alpha_E(y_i)-\alpha_F(y_i)|}{2}\right)\pi+\sin\left(\frac{|\beta_E(y_i)-\beta_F(y)|}{2}\right)\pi\right]=D_{\mathrm{IF}_{S1}}(E,F)$$

Similarly, we can prove for eq. (4.10).

(ii) Let us consider for

$$D_{\mathrm{IF}_{S1}}(E,G)+D_{\mathrm{IF}_{S1}}(F,G)-D_{\mathrm{IF}_{S1}}(E\cup F,G)$$
$$=\frac{1}{2n}\sum_{i=1}^{n}\left[\sin\left(\frac{|\alpha_E(y_i)-\alpha_G(y_i)|}{2}\right)\pi+\sin\left(\frac{|\beta_E(y_i)-\beta_G(Y_i)|}{2}\right)\pi\right]$$
$$+\frac{1}{2n}\sum_{i=1}^{n}\left[\sin\left(\frac{|\alpha_F(y_i)-\alpha_G(y_i)|}{2}\right)\pi+\sin\left(\frac{|\beta_F(y_i)-\beta_G(y_i)|}{2}\right)\pi\right]$$
$$-\frac{1}{2n}\sum_{i=1}^{n}\left[\sin\left(\frac{|\alpha_{E\cup F}(y_i)-\alpha_G(y_i)|}{2}\right)\pi+\sin\left(\frac{|\beta_{E\cup G}(y_i)-\beta_G(y_i)|}{2}\right)\pi\right]$$
$$=\frac{1}{2n}\sum_{i=1}^{n}\left[\left(\frac{|\alpha_E(y_i)-\alpha_G(y_i)|}{2}\right)\pi+\sin\left(\frac{|\beta_E(y_i)-\beta_G(y_i)|}{2}\right)\pi\right]$$
$$+\frac{1}{2n}\sum_{i=1}^{n}\left[\sin\left(\frac{|\alpha_F(y_i)-\alpha_G(y_i)|}{2}\right)\pi+\sin\left(\frac{|\beta_F(y_i)-\beta_G(y_i)|}{2}\right)\pi\right]$$

$$-\frac{1}{2n}\left[\sum_{y_1\in Y_1}\sin\left(\frac{|\alpha_F(y_i)\ -\alpha_G(y_i)|}{2}\right)\pi+\sin\left(\frac{|\beta_F(y_i)\ -\beta_G(y_i)|}{2}\right)\pi\right]$$

$$+\sum_{y_1\in Y_2}\left[\sin\left(\frac{|\alpha_E(y_i)\ -\alpha_G(y_i)|}{2}\right)\pi+\sin\left(\frac{|\beta_E(y_i)\ -\beta_G(y_i)|}{2}\right)\pi\right]$$

$$=\frac{1}{2n}\left[\sum_{y_1\in Y_1}\sin\left(\frac{|\alpha_E(y_i)\ -\alpha_G(y_i)|}{2}\right)\pi+\sin\left(\frac{|\beta_E(y_i)\ -\beta_G(y_i)|}{2}\right)\pi\right]$$

$$+\frac{1}{2n}\sum_{y_1\in Y_2}\left[\sin\left(\frac{|\alpha_F(y_i)\ -\alpha_G(y_i)|}{2}\right)\pi+\sin\left(\frac{|\beta_F(y_i)\ -\beta_G(y_i)|}{2}\right)\pi\right]\geq 0$$

$$\Rightarrow D_{\mathrm{IF}_{S1}}(E,G)+D_{\mathrm{IF}_{S1}}(F,G)-D_{\mathrm{IF}_{S1}}(E\cup F,G)\geq 0$$

Similarly, we can prove

$$D_{\mathrm{IF}_{S2}}(E,G)+D_{\mathrm{IF}_{S2}}(F,G)-D_{\mathrm{IF}_{S2}}(E\cup F,G)\geq 0$$

(iii)

$$D_{\mathrm{IF}_{S1}}(E,G)+D_{\mathrm{IF}_{S1}}(F,G)\ -D_{\mathrm{IF}_{S1}}(E\cap F,G)\geq 0$$

and

$$D_{\mathrm{IF}_{S2}}(E,G)+D_{\mathrm{IF}_{S2}}(F,G)-D_{\mathrm{IF}_{S2}}(E\cap F,G)\geq 0\quad\forall E,F,G\in\mathrm{IF_S}$$

4.4 Cosine trigonometric intuitionistic fuzzy similarity measures

Definition 4.9. Let E and F be two IFS on universal set Y as $Y=\{y_1,y_2,\ldots,y_n\}$, whose membership values are $\alpha_E(y_i),\ i=1,2,\ldots,n$ and $\alpha_F(y_i),\ i=1,2,\ldots,n$, respectively. Then the measures of cosine intuitionistic fuzzy similarity $S_{\mathrm{IF_M}}(E,F)$ between $S_{\mathrm{IF_S}}(E,F)$ fuzzy sets E and F are defined as

$$S_{\mathrm{IF}_{S1}}(E,F)=\frac{1}{2n}\sum_{i=1}^{n}\left[\cos\left(\frac{|\alpha_E(y_i)\ -\alpha_F(y_i)|}{2}\right)\pi+\cos\left(\frac{|\beta_E(y_i)\ -\beta_F(y_i)\,|}{2}\right)\pi\right]\tag{4.15}$$

$$S_{\mathrm{IF}_{S2}}(E,F)=\frac{1}{2n}\sum_{i=1}^{n}\left[\cos\left(\frac{|\sqrt{\alpha_E(y_i)}-\sqrt{\alpha_F(y_i)}|}{2}\right)\pi+\cos\left(\frac{|\sqrt{\beta_E(y_i)}-\sqrt{\beta_F(y_i)}\,|}{2}\right)\pi\right]\tag{4.16}$$

Theorem 4.2. *Identities* (4.15) *and* (4.16) *prove all the conditions of intuitionistic fuzzy similarity measures.*

Proof. Given

$$S_{\mathrm{IF_{S1}}}(E,F)=\frac{1}{2n}\sum_{i=1}^{n}\left[\cos\left(\frac{|\alpha_E(y_i)\ -\alpha_F(y_i)|}{2}\right)\pi+\ \cos\left(\frac{|\beta_E(y_i)\ -\beta_F(y_i)\,|}{2}\right)\pi\right]$$

Here we prove all four conditions of intuitionistic fuzzy distance measures that are mentioned in Definition 4.7:

$$\text{(IS1)}\ S_{\mathrm{IF_{S1}}}(E,F)=\frac{1}{2n}\left[\sum_{i=1}^{n}\cos\left(\frac{|\alpha_E(y_i)-\alpha_F(y_i)|}{2}\right)\pi+\cos\left(\frac{|\beta_E(y_i)-\beta_F(y_i)|}{2}\right)\pi\right]$$

$$=\frac{1}{2n}\left[\sum_{i=1}^{n}\cos\left(\frac{|-(\alpha_F(y_i)-\beta_E(y_i))|}{2}\right)\pi+\cos\left(\frac{|-(\beta_F(y_i)-\beta_E(y_i))|}{2}\right)\pi\right]$$

(IS2) $S(d,d^C)=0,\text{if}\ d\,\text{is the crisp set}$

(IS3) $S(K,K)=\max_{E,F\in\mathrm{IF_S}}D(E,F)\ \ \forall K\in\mathrm{IF_S}$

(IS4) $\forall E,F,G\in\mathrm{IF_S},\ \text{if}\ E\subset F\subset G\ \text{then}\ \alpha_E(y_i)\le\alpha_F(y_i)\le\alpha_G(x_i)$ and

$$\beta_E(y_i)\le\beta_F(y_i)\le\beta_G(y_i)=\frac{1}{2n}\left[\left(\frac{|(\alpha_E(y_i))-\alpha_F(y_i))|}{2}\right)\pi+\left(\frac{|(\beta_E(y_i)-\beta_F(y_i))|}{2}\right)\pi\right]$$

$$\ge\frac{1}{2n}\left[\left(\frac{|(\alpha_E(y_i)-\alpha_G(y_i))|}{2}\right)\pi+\left(\frac{|(\beta_E(y_i)-\beta_G(y_i))|}{2}\right)\pi\right]$$

and the nature of cosine function is given by

$$\frac{1}{2n}\left[\sum_{i=1}^{n}\cos\left(\frac{|(\alpha_E(y_i)-\alpha_G(y_i))|}{2}\right)\pi+\cos\left(\frac{|(\beta_E(y_i)-\beta_G(y_i))|}{2}\right)\pi\right]$$

$$\le\frac{1}{2n}\left[\sum_{i=1}^{n}\cos\left(\frac{|(\alpha_E(y_i)-\alpha_F(y_i))|}{2}\right)\pi+\cos\left(\frac{|(\beta_E(y_i)-\beta_F(y_i))|}{2}\right)\pi\right]$$

$$S_{\mathrm{IF_{S1}}}(E,F)\ge S_{\mathrm{IF_{S1}}}(E,G)$$

Similarly,

$$\frac{1}{2n}\left[\left(\frac{|(\alpha_E(y_i)-\alpha_G(y_i))|}{2}\right)\pi+\left(\frac{|(\beta_E(y_i)-\beta_G(y_i))|}{2}\right)\pi\right]$$

$$\ge\frac{1}{2n}\left[\left(\frac{|(\alpha_F(y_i)-\alpha_G(y_i))|}{2}\right)\pi+\left(\frac{|(\beta_F(y_i)-\beta_G(y_i))|}{2}\right)\pi\right]$$

and the nature of cosine function is given by

$$\frac{1}{2n}\left[\sum_{i=1}^{n}\cos\left(\frac{|(\alpha_E(y_i)-\alpha_G(y_i))|}{2}\right)\pi+\cos\left(\frac{|(\beta_E(y_i)-\beta_G(y_i))|}{2}\right)\pi\right]$$

$$\leq\frac{1}{2n}\left[\sum_{i=1}^{n}\cos\left(\frac{|(\alpha_F(y_i)-\alpha_G(y_i))|}{2}\right)\pi+\cos\left(\frac{|(\beta_F(y_i)-\beta_G(y_i))|}{2}\right)\pi\right]$$

$$S_{\mathrm{IF_{S1}}}(E,F)\geq S_{\mathrm{IF_{S1}}}(F,G)$$

Similarly, we can prove the identity given by eq. (4.16).

4.5 Properties of cosine trigonometric intuitionistic fuzzy similarity measures

Let T be a similarity measure on IFS and for $E,F\in \mathrm{IF_S}$. If they satisfy that for all $y_i\in Y$, either $E(y_i)\subseteq F(y_i)$ *or* $E(y_i)\supseteq F(y_i)$, then

$$\text{(i) } T_{\mathrm{IF_{S1}}}(E\cup F,E\cap F)=T_{\mathrm{IF_{S1}}}(E,F) \tag{4.17}$$

and

$$T_{\mathrm{IF_{S2}}}(E\cup F,E\cap F)=T_{\mathrm{IF_{S2}}}(E,F)\quad\forall E,F,G\in\mathrm{IF_S} \tag{4.18}$$

and

$$\text{(ii) } T_{\mathrm{IF_{S1}}}(E\cup F,G)\leq T_{\mathrm{IF_{S1}}}(E,G)+T_{\mathrm{IF_{S1}}}(F,G) \tag{4.19}$$

$$T_{\mathrm{IF_{S2}}}(E\cup F,G)\leq T_{\mathrm{IF_{S2}}}(E,G)+T_{\mathrm{IF_{S2}}}(F,G)\quad\forall E,F,G\in\mathrm{IF_S} \tag{4.20}$$

$$\text{(iii) } T_{\mathrm{IF_{S1}}}(E\cap F,G)\leq T_{\mathrm{IF_{S1}}}(E,G)+T_{\mathrm{IF_{S1}}}(F,G) \tag{4.21}$$

and

$$T_{\mathrm{IF_{S2}}}(E\cap F,G)\leq T_{\mathrm{IF_{S1}}}(E,G)+T_{\mathrm{IF_{S2}}}(F,G)\quad\forall E,F,G\in\mathrm{IF_S} \tag{4.22}$$

Proof.

(i) Using Definition 4.7

$$T_{\mathrm{IF_{S1}}}(E\cup F,E\cap F)=\frac{1}{2n}\sum_{i=1}^{n}\left[\cos\left(\frac{|\alpha_{E\cup F}(y_i)\ -\alpha_{E\cap F}(y_i)|}{2}\right)\pi\right.$$

$$\left.+\cos\left(\frac{|\beta_{E\cup F}(y_i)\ -\beta_{E\cap F}(y_i)|}{2}\right)\right]$$

$$= \frac{1}{2n}\left[\sum_{y_i \in Y_1}\left[\cos\left(\frac{|\alpha_F(y_i) - \alpha_E(y_i)|}{2}\right)\pi + \cos\left(\frac{|\beta_F(y_i) - \beta_E(y_i)|}{2}\right)\pi\right]\right.$$

$$\left. + \sum_{y_i \in Y_2}\left[\cos\left(\frac{|\alpha_E(y_i) - \alpha_F(y_i)|}{2}\right)\pi + \cos\left(\frac{|\beta_E(y_i) - \beta_F(y_i)|}{2}\right)\pi\right]\right]$$

$$= \frac{1}{2n}\sum_{i=1}^{n}\left[\cos\left(\frac{|\alpha_E(y_i) - \beta_F(y_i)|}{2}\right)\pi + \cos\left(\frac{|\beta_E(y_i) - \beta_F(y_i)|}{2}\right)\pi\right]\Bigg]$$

$$= T_{\mathrm{IF_{S1}}}(E \cup F, E \cap F)$$

Similarly we can prove eq. (4.18).

(ii) Let us consider for

$$T_{\mathrm{IF_{S1}}}(E, G) + T_{\mathrm{IF_{S1}}}(F, G) - T_{\mathrm{IF_{S1}}}(E \cup F, G)$$

$$= \frac{1}{2n}\sum_{i=1}^{n}\left[\cos\left(\frac{|\alpha_E(y_i) - \alpha_G(y_i)|}{2}\right)\pi + \cos\left(\frac{|\beta_E(y_i) - \beta_G(y_i)|}{2}\right)\pi\right]$$

$$+ \frac{1}{2n}\sum_{i=1}^{n}\left[\cos\left(\frac{|\alpha_F(y_i) - \alpha_G(y_i)|}{2}\right)\pi + \cos\left(\frac{|\beta_F(y_i) - \beta_G(y_i)|}{2}\right)\pi\right]$$

$$- \frac{1}{2n}\sum_{i=1}^{n}\left[\cos\left(\frac{|\alpha_{E\cup F}(y_i) - \alpha_G(y_i)|}{2}\right)\pi + \cos\left(\frac{|\beta_{E\cup G}(y_i) - \beta_G(y_i)|}{2}\right)\pi\right]$$

$$= \frac{1}{2n}\sum_{i=1}^{n}\left[\cos\left(\frac{|\alpha_E(y_i) - \alpha_G(y_i)|}{2}\right)\pi + \cos\left(\frac{|\beta_E(y_i) - \beta_G(y_i)|}{2}\right)\pi\right]$$

$$+ \frac{1}{2n}\sum_{i=1}^{n}\left[\cos\left(\frac{|\alpha_F(y_i) - \alpha_G(y_i)|}{2}\right)\pi + \cos\left(\frac{|\beta_F(y_i) - \beta_G(y_i)|}{2}\right)\pi\right]$$

$$= \frac{1}{2n}\left[\sum_{y_1 \in Y_1}\cos\left(\frac{|\alpha_E(y_i) - \alpha_G(y_i)|}{2}\right)\pi + \cos\left(\frac{|\beta_E(y_i) - \alpha_G(y_i)|}{2}\right)\pi\right]$$

$$+ \sum_{y_1 \in Y_2}\left[\cos\left(\frac{|\alpha_E(y_i) - \alpha_G(y_i)|}{2}\right)\pi + \cos\left(\frac{|\beta_E(y_i) - \beta_G(y_i)|}{2}\right)\pi\right]$$

Similarly, we can prove $T_{\mathrm{IF_{S2}}}(E \cup F, G) \le T_{\mathrm{IF_{S2}}}(E, F) + T_{\mathrm{IF_{S2}}}(F, G) \quad \forall E, F, G \in \mathrm{IF_S}$

(iii) Similarly, we can prove

$$T_{\mathrm{IF_{S1}}}(E \cap F, G) \le T_{\mathrm{IF_{S1}}}(E, G) + T_{\mathrm{IF_{S1}}}(F, G)$$

and

$$T_{\mathrm{IF_{S1}}}(E \cap F, G) \le T_{\mathrm{IF_{S1}}}(E, G) + T_{\mathrm{IF_{S1}}}(F, G) \quad \forall E, F, G \in \mathrm{IF_S}$$

4.6 Conclusion

First, we defined IFS. Then, we introduced a general intuitionistic inclusion on a universe of discourse X and intuitionistic similarity measure between two IFS on X. Definitions have been given to illustrate the concept of IFS, intuitionistic similarity between two IFS, and intuitionistic fuzzy relation based on intuitionistic similarity between two IFS, intuitionistic fuzzy distance measure, sine and cosine trigonometric intuitionistic fuzzy distance measures. Properties of sine and cosine trigonometric intuitionistic fuzzy distance measures are also given.

Disclosure statement

No potential conflict of interest was reported by the authors.

References

[1] Zadeh L.A. (1965) Fuzzy sets, Information and Control, 8, 338–356.
[2] Atanassov K. (1986) Intuitionistic fuzzy sets, Fuzzy Sets and Systems, 20, 87–96.
[3] Atanassov K. & Gargov G. (1990) Intuitionistic fuzzy logic, Comptes Rendus de l' Academic Bulgare des Sciences, 43(3), 9–16.
[4] Atanassov K. Intuitionistic fuzzy sets: Theory and applications. Physica-Verlag, Heidelberg, 1999.
[5] Burillo P. & Bustince H. (1996) Entropy on intuitionistic fuzzy sets and interval-valued fuzzy sets, Fuzzy Sets and Systems, 78, 305–316.
[6] Atanassov K. (1994) New operations defined over the intuitionistic fuzzy sets, Fuzzy Sets and Systems, 61, 137–142.
[7] Liu X. (1992) Entropy, distance measure and similarity measure of fuzzy sets and their relations, Fuzzy Sets and Systems, 52, 305–318.

Alok Dhaundiyal, Suraj B. Singh

5 Distributed Activation Energy Modeling by the transmutation of different density functions

Abstract: This chapter incorporates the different perspective of modeling of activation energies through ingression of an additional parameter of distribution function to increase the flexibility and controlling ability of modeling by the linear mixing. The transmuted functions are adopted to bring flexibility in the numerical solution of multireaction models. A stochastic model is used to represent the activation energy function. Thus, the different activation energies can be represented by a density function to describe the pyrolysis process. The qualitative, as well as quantitative effects of the transmuted family of distribution on biomass pyrolysis, are carefully examined by the proposed mathematical solution. The nonlinear thermal history is implemented to demonstrate the approach.

Keywords: Biomass pyrolysis, transmuted function, kinetic parameters, thermal parameters, nonlinear ramping

Notation

"*t*"	Transmuted form	Dimensionless
W	Weight parameter for volatile content	Dimensionless
T	Temperature	K
α	Conversion	Dimensionless
A	Frequency factor	min^{-1}
E	Activation energy	kJ/mol
$p(E)$	Density function of activation energy	Dimensionless
τ	Shape parameter	Dimensionless
R	Universal gas constant	kJ/K mol
θ	Shape parameter for the Lindley distribution	Dimensionless
ψ	Shape parameter for the Lindley distribution	Dimensionless

Alok Dhaundiyal, Szent István University, Institute of Process Engineering, Gödöllő, Hungary
Suraj B. Singh, Department of Mathematics, Statistics and Computer Science, Govind Ballabh Pant University of Agriculture and Technology, Pantnagar, Uttarakhand, India

https://doi.org/10.1515/9783110628616-005

γ	Location parameter for the Lindley distribution	kJ/mol
β	Shape parameter for the Weibull distribution	Dimensionless
η	Scale parameter for the Weibull distribution	kJ/mol
γ	Location parameter of the Weibull distribution	kJ/mol
ρ	Shape parameter of the Frechet distribution	Dimensionless
δ	Scale parameter of the Frechet distribution	kJ/mol
μ	Location parameter of the Frechet distribution	kJ/mol
E_0	Mean value of the Gaussian distribution	kJ/mol
σ	Standard deviation of the Gaussian distribution	kJ/mol
ϕ	Location parameter of the Rayleigh distribution	kJ/mol
χ	Scale parameter of the Rayleigh distribution	kJ/mol
ε	Stochastic nature of regression model (error)	Dimensionless
a	Rate of change of heating rate	$°C/min^2$
s	Instantaneous value of time interval	Min
m	*Heating rate*	$°C/min$
T_0	Initial temperature	K
T_f	The maximum temperature	K
E_s	Central value of activation energy	kJ/mol
E_w	Step width	kJ/mol
$W(\kappa)$	Lambert *W* function	Dimensionless
κ	Time rescaling factor	Dimensionless
y^e	Maxima point	Dimensionless
y_s	Energy rescaling corresponding central value	Dimensionless
y_w	Energy rescaling factor for step width	Dimensionless
ω	Dimensionless number	Dimensionless
E_∞	Upper bound of activation energy	kJ/mol
f_1 and f_2	The components of the Lindley distribution	Dimensionless

5.1 Introduction

Modeling of a system is one of the most challenging tasks when you are dealing with the unit where multivariables are acting simultaneously within a system. One of such systems is a biological activity of constituents that have different chemical, thermal, and physical properties, and it emerges out to be a reason of anomalies in the measurands. There are numerous ways of tackling such complicated problems, which need myriads of steps to evaluate them properly. Therefore, some assumptions, approximations, and advanced calculus are required to save three M's (men, machine, and money) from overexploitation.

Here, one of the thermochemical processes, namely pyrolysis, has been analyzed qualitatively and quantitatively of a given problem. The problem arises when the dynamical behavior of inflexion point of thermogravimetric (TG) variation of mass curve impedes the predicted stochastic modeling of activation energies to converge around the neighboring point of it. Many methodologies [1–7] have been adopted to fetch the best suitable prediction that can converge up to 90% of the measurand value.

Numerical solution of multireaction modeling of parallel reactions provides a better insight of parametric behavior and its effect on the natural phenomenon; thereafter, it becomes a lot easier to predict the outcome of solution than that of statistical linear models that merely rely on the linear regression of a straight line. There are various methods of modeling and evaluating kinetic parameters of pyrolysis process that can also be categorized as the single reaction model and the multireaction model [8–12].

The demarcation of modeling is based on the end results required by the user; therefore, it is necessary to understand the requirement of the given problem. In the same context, exploratory and application-based models are of the main concern [13]. The exploratory models require less computation resources and it is relatively fast and flexible, but the accuracy is highly compromised. On the other hand, the application-based models look for the concreate, robust, mature, and precise solution. It requires a little computer-based programming, but solution converges without reservation. In the real-time problems, the application-based modeling provides an accurate and sophisticated numerical solution which can be delineated graphically.

Hence, in this study the effect of one of such application-based models is considered to carry out the simulation process. Thermochemical parameters along with distribution pattern are examined through modeling of activation energies by the transmuted functions. Moreover, the dynamic behavior of stage decomposition is depicted through nonstationary maxima of curve with respect to time. The purpose of study is solely based on the application perspective of transmuted function, reducing the variation gap between Arrhenius parameters, and proposing the new concept for modeling activation energies, not to over-fitting of curves. Therefore, a new scheme has been proposed where the kinetic parameters are not influenced by the inflexion point of two different reaction phases, which is overlooked in the previous studies [1, 3, 14].

5.2 The transmuted function of activation energy *p*(*E*)

The idea of grouping the parallel reactions to analyze the mechanism of complex reactions is not the new scheme of comprehending the interrelationship between frequency factor and activation energy. For surface-catalyzed reactions, it had been envisaged that the surface has various sites that rely on distribution of activation energies among the competitive reactions [15]. The similar concept was adopted to measure irreversible changes in an electrical resistance while annealing the evaporated metal films [16]. This scheme was highly adopted by the coal pyrolysis community and the parallel reactions approach came into light in the 1960s [17, 18]. Later, Anthony and Howard (1976) proposed that the activation energy of the parallel reactions could be expressed as a density function of the normal distribution for coal pyrolysis [19]. However, there are some asymmetrical functions tested to validate the result for the loose biomass [20–22]. It is not that the derived results from other distribution functions have higher degree of entropy than the normal distribution. But the compensatory effect due to the mathematical mandate in the old theories that get drifted off through modification of the prevailing theorem [1].

Despite all those efforts for calculating multivariate function [1, 23], still there is a scope of assessing the nature of stochastic function as well as selection of appropriate thermal programming for pyrolysis reactions. Although it is difficult to interpret TG and derivative TG curves exactly, since (i) it is impossible to measure the sample temperature exactly through thermocouple, (ii) there is a time lag while formation of intermediate stages, (iii) interference of interactive atmosphere or cooling effect, and (iv) heat transfer limitation. Experimental curves have constraints and stagnation with time; overall, it reflects the gradual variation in property of material, which is related to alteration of thermal history. On the other hand, the attribute of mathematical model is based on the set protocol that overlooks the miscellaneous factors related to thermal decomposition. Transformation time between intermediate stage is almost Δt, which tends to zero for the mathematical model.

There are some methodologies developed to simulate the TG attribute effectively such as double-distributed activation energy model (2-DAEM) [5, 14]. The scheme manifests to mitigate uncertainties of distribution function to describe TG curves by introducing the weight parameter for volatile content "w" [24]. This proposition can be valid unless pyrolysis of biomass is carrying out in the unrealistic condition since it is not necessary that the role of volatile content ceases to act with the decreasing value of w. Lewellen et al. [25] stated that the residence time of volatile content has significant influence on pyrolysis of biomass. It decides the extent of secondary pyrolysis. Therefore, the weight factor and time factor cannot be correlated, as competitive pathways such as cracking, repolymerization can be inhibited through formation of some of light volatile gases. The volatile content may initiate the autocatalytic reactions at elevated temperature. It accelerates the formation of the stable products content which would escape the biomass matrix to yield volatiles [25]. There are various key issues related to modeling of biomass that have been summarized in some of literatures [25–28]. It is complicating to find the exact mathematical solution to interpret the TG curves, but the discordance can be subdued to some extent. There are some other mathematical solutions used for condensed phase reactions – isoconversional and model fitting. Both are highly recommended for kinetic analysis [13, 29].

Regression-based modeling is very promising and time-saving, whether the estimated results are precise and exhibit the kinetic mechanism of pyrolysis. This must be investigated carefully. Isoconversional and model fitting are some of regression-based methods used for estimating the kinetic parameters. Isoconversional scheme or model-free kinetics does not imbibe the idea of fitting the experimental TG data to the phenomenological models, but it does not infer "free from assumption." The basic tenets of scheme are that at any given extent of reaction, the same reactions occur in the same ratio at different heating rates and the reactions must be independent of temperature. However, this hypothesis can be easily nullified by a system of parallel independent reactions whose relative reactivity is proportional to temperature and it is varying due to different activation energies [30]. It is also possible that the overall reaction pathway of the competitive reactions with different activation energies is different at different temperature [31], as Lewellen et al. [25] suggested for pyrolysis of cellulose. Unlike isoconversional method, inconsistency for determining Arrhenius parameters is observed in the model-fitting method. It is mainly due to force-fitting of date to the hypothetical reaction model. The Arrhenius parameters are determined by the phenomenological model, which is already assumed. Moreover, both temperature and conversion consistently vary with time. This methodology is unable to demarcate separately the temperature dependence of rate constant and the conversion. Eventually, it causes drastic variation while determining the frequency factor and activation energies, and any phenomenological model can be claimed to describe the pyrolysis mechanism of biomass. Moreover, the existence of competitive reactions with different activation energies affects the overall decomposition rate, as these reactions are highly influenced by temperature and extent of conversion; therefore, the activation calculated from this method is a function of T and α [29]. This model is highly inaccurate and has been highly discouraged for kinetic analysis [32].

However, instrumental noise, response time, inabilities of thermocouples, and sensitivity of thermobalance cannot be incorporated in any advance integral model or regression schemes. One can filter data to remove suspicious values and reduce the gap between A and E variation. One may also adopt statistical testing scheme (Fisher's exact test, Dixon Q's test, Bartlett's test, etc.). But if your thermoanalytical data size is colossal, these probing schemes become time-consuming to analyze the sample size. Distribution in activation energies through conventional logistical distribution function [33] is not only the way to introduce a reactivity distribution, but there are several other ways to solve DAEM, such as linear mixing of density function $p(E)$ of activation energy:

$$p^t(E) = (1+\tau)p(E) - \tau\ (p(E))^2,\ |\tau|\ \leq 1 \tag{5.1}$$

where E is the activation energy.

Unknown activation energy of pyrolysis reaction is said to be transmuted if it follows eq. (5.1). However, the same expression holds valid for cumulative distribution function of activation energy. Parametrization of well-established distribution reduces the effective computation time for deriving more robust and flexible families of distribution. Shaw and Buckle [34] were the leading researchers to explore distributional flexibility through introducing a new parameter to an existing distribution. They figured out that involvement of parameters did not make solution cumbersome but enhance the reproducibility of the modeling of natural process. The generated or transmuted family imbibes the parent distribution and impart more dynamical attribute to simulated results for various types of data than the conventional logistical approach. Implementation of transmuted family is very vast, and it encompasses all the natural phenomenon that demands the high degree of flexibility as the entropy of such systems is very high. One of such study is carried out for modeling the metrological data. The transmuted Gumbel distribution is found to be very effective for analyzing the snowfall data of a region [35]. The competitiveness of transmuted exponentiated Frechet distribution over exponentiated Frechet distribution for modeling wind speed data is performed to gather the statistical properties for a wind energy conversion system. The application indicates that the transmuted model's performance is relatively good to the parent model [36]. Another field of application of transmuted function is the mortality or failure of some system with respect to time. It is found that the reliable behavior of transmuted distribution functions is very useful for modeling reliability data in the manufacturing industries [37]. The application as well as modification of the conventional distributions is reported in the several literatures [38–40].

In the subsequent sections, the approximation techniques are adopted to simplify the classical DAEM and the conventional density variable from transmuted family of distribution, so that the distribution of reactivity can be introduced through the proposed scheme. However, interdependence of frequency factor (A) and E will still prevail in modeling of DAEM, but it is assumed that frequency factors (A) for all the competitive reactions are invariant and assumed to be constant. DAEM is represented through double integral equations for first-order reaction (eq. (5.2) [1]:

$$1 - \alpha = \int_0^\infty \exp\left\{\left(-\int_0^t Ae^{-\frac{E}{RT}}dt\right)\right\}p(E)dE \tag{5.2}$$

Density and cumulative functions of activation energy are illustrated in Table 5.1.

Table 5.1: Distribution functions and their corresponding parameters.

Distribution	Density function $p(E)$	Cumulative function $G(E)$	$\xi > 0$ Parameters
Lindley [41]	$\frac{\theta^2}{\theta+\psi}\left(1+\psi(E-\gamma)\right)e^{-\theta(E-\gamma)}$	$\left\{1-\left(1+\frac{\theta\psi(E-\gamma)}{\theta+\psi}\right)e^{-\theta(E-\gamma)}\right\}$	$\theta,\ \psi,\gamma$
Weibull [42]	$\frac{\beta}{\eta}\left(\frac{E-\gamma}{\eta}\right)^{(\beta-1)}e^{-\left(\frac{E-\gamma}{\eta}\right)^{\beta}}$	$\left\{1-e^{-\left(\frac{E-\gamma}{\eta}\right)^{\beta}}\right\}$	$\beta,\ \eta,\ \gamma$
Frechet [43]	$\frac{\rho}{\delta}\left(\frac{E-\mu}{\delta}\right)^{-(1+\rho)}e^{-\left(\frac{E-\mu}{\delta}\right)^{-\rho}}$	$\left\{e^{-\left(\frac{E-\mu}{\delta}\right)^{-\rho}}\right\}$	$\rho,\ \delta,\ \mu$
Gaussian [44]	$\frac{1}{\sqrt{2\pi\sigma^2}}e^{-\left(\frac{E-E_0}{\sqrt{2}\sigma}\right)^2}$	$\left\{\frac{1}{2}\left(1+\text{erf}\left(\frac{E-E_0}{\sigma\sqrt{2}}\right)\right)\right\}$	$E_0,\ \sigma$
Rayleigh [45]	$2\chi(E-\phi)\exp\left\{-\chi(E-\phi)^2\right\}$	$1-\exp\left\{-\chi(E-\phi)^2\right\}$	ϕ,χ

5.2.1 Distributed activation energy modeling

A complex material such as biomass or fossil fuel comprises several constituents whose thermal decomposition varies greatly with time and temperature. The different products are generated at different stages of decomposition. If the parallel reactions follow the complex pathways, it becomes more complicating to model such process, so it is possible that a simple mathematical distribution will not have flexibility to describe the reaction profile of some reactions. Copula-based modeling [1, 23] is recently adopted to illustrate the complexity of decomposition reactions and to some extent it is found to be very helpful; however, the validation of results for different thermal histories is not done yet. Discrete activation energy distributions schemes are developed to provide flexibility to the reaction model. One of such models was developed for petroleum generation, which was first published by Ungerer and Pelet [46]. This scheme is derived from the isoconversional methods, which is mainly designed for estimating the unique values of Arrhenius parameters, but later it was used to correlate activation energy variation with conversion. The detailed study of development of several other models is reported in the literature [47]. But it requires special attention for energy spacing between reaction channels, which may affect the validity of the model. Moreover, it is purely based on the non-linear regression and the numerical integration, which itself is stochastic in nature. A regression analysis is based on the relationship between study and explanatory variables through the series of coefficients that decide the characteristics and role of independent variables in any event:

$$y = f(X_1,\ X_2,\ \ldots, X_k,\ a_1,\ a_2,\ \ldots, a_k) + \varepsilon \tag{5.3}$$

where ε represents the stochastic nature of regression model that can alter the solution if activation energy is not discretized evenly. However, this scheme is highly encouraged for estimating the kinetic properties of kerogen pyrolysis rates [13, 48]. A major drawback in discrete modeling is the derivation of the structure, that is, of the terms in the discrete-activation energy model. When the form of differential equation generating data is identified, one can estimate possible terms for the discrete model from numerical methods, but not necessarily the associated coefficients. A good structure design (i.e., the form of $f(X_1,\ X_2,\ \ldots\ ,X_k,\ a_1,\ a_2,\ \ldots a_k)$) gets the kinetic model–fitted data properly and provide the good predictive results. Redundant terms in a model, on the one hand, can lead to over fitting, whereas too few terms make model ineffective. There are some practical criteria considered for effective modeling. First, the model should be numerically stable. Second, the selected model makes the best predictions, which can be performed by testing the given data for different heating rates or thermal profile. Asymptotic behavior of model should comply with those of the data. Keep the demarcation of discrete and continuous models for abeyance and formulate the transmuted function-based scheme.

To demonstrate the scheme, eq. (5.2) is bifurcated for simplifying the mathematical expression of DAEM. It comprises two parts: double exponential term (DExp) and the continuous density function of activation energy. Laplace integral method is implemented to estimate the approximated form of DAEM. DExp terms are an implicit function of time and temperature history experienced by the sample. The second part is independent of time and depends on the distribution of reactivity. First, the behavior of implicit function is perused, and then approximations are derived for the nonlinear profile.

The ideal thermal histories are seldom implemented for the industrial applications; therefore, the computational algorithm is tested for the nonlinear thermal profile represented as follows (eq. (5.4)):

$$T(K) = as^2 + T_0(K) \tag{5.4}$$

where $a\left(\frac{°\mathrm{C}}{\mathrm{min}^2}\right) = \frac{dm}{dt}$ and the value of "a" represents the rate of change of heating rate (m) with time.

Let, $f(T) = -\frac{E}{RT}$

$$\mathrm{DExp} = \exp\left\{\left(-\int_{T_0}^{T_f} \frac{A}{2a\sqrt{\left(\frac{T-T_0}{a}\right)}} \exp(f(T_f))dT\right)\right\} \tag{5.5}$$

The ratio of $\frac{E}{R(as^2+T_0)} \to \infty$ is assumed to be large; thus, the significant contribution of integral is at neighborhood of $T - T_f$ and $f(T_f) > 0$. This provides the asymptotic approximation to the temperature-dependent function:

$$\text{DExp}\sim\exp\left(\frac{A}{2\sqrt{a(T_f - T_O)}}\frac{-\exp(f(T_f))}{f'(T_f)}\right) \tag{5.6}$$

Correlating the result with the central value and the step size of activation energies, eq. (5.6) can be rewritten as

$$\text{DExp}\sim\exp\left\{-\exp\left(\frac{E_s - E}{E_w}\right)\right\} \tag{5.7}$$

Rather than arbitrarily spacing the activation energies, it is better to generate the form of function that relies on the explanatory variables of experiments, as the trend of activation energy variation with respect to conversion is highly indeterministic [29].

Let, $g(E) = \left(\frac{E_S - E}{E_W}\right)$.

Asymptotic approximation results in a rapidly varying function as $\frac{E}{R(as^2+T_O)} \to \infty$ and function varies from zero to one as E increases by E_w around E_s. To further simplify the function, $\left(\frac{E_S - E}{E_W}\right)$ is to be expanded by the Taylor series about $E = E_S$:

$$g(E)\sim g(E_s) + (E - E_s)g'(E_s) + \text{L}\cdots \tag{5.8}$$

The initial condition for function $g(E)$ at E_s is defined as

$$g(E_s) = 0 \text{ and } g'(E_s) = -\frac{1}{E_W}.$$

Equating eqs. (5.6) and (5.7), we have

$$g(E) \equiv -f(T_f) + \ln\left(\frac{A}{2f'(T_f)\sqrt{a(T_f - T_O)}}\right) \tag{5.9}$$

After solving eqs. (5.8) and (5.9), we have

$$E_s = RT_fW(\kappa)$$

$$E_W = \frac{RE_s(T_f)}{E_s + RT_f}$$

where both E_s and E_W are implicit functions of time. W is a Lambert W function and the dimensionless group $\left(\frac{T_f A}{2at}\right)$ is denoted by κ.

Note: Lambert W function is defined as one of the real roots of equation of form

$$xe^x = W(x)$$

Approximations of Lambert *W* function for small and large values of "*x*" (correlated to short and long durations of time) can be represented as

$$W \sim x - x^2, \ x = 1$$

$$W \sim \ln\left(\frac{x}{\ln\left(\frac{x}{\ln(x)}\right)}\right) \ 1 = x$$

5.2.2 Characteristic of distribution function

It is observed from the approximated form of DExp, correlated with activation through energy spacing of step size E_w, that the function behaves like a smoothed step function. It rises swiftly from zero to one as the term $\left(\frac{A}{2\sqrt{a(T_f - T_0)}}\right)$ increases, or $\left(\frac{E - E_S}{E_W}\right) \rightarrow \infty$ and the peak is concentrated at the neighborhood of $T = T_f$. The DExp as implicit function *E* makes the whole integrand homogenous and it becomes easier to compute the complicated integral problem. But the fact that cannot be overshadowed is the relative significance of E_w to the standard deviation of activation energies from the average value of activation energy. In eq. (5.2), the DExp is multiplied with the density function of activation energy. The mean value as well as standard deviation of distribution function is constant; therefore, the shape of the whole integrand relies on the study variables E_w and E_s. There must be the limits that can modulate the form of the model through changing the location of maximum of the total integrand. Thus, it provides two different ways of solving DAEM. The first case: when the standard deviations of continuous distribution functions (listed in Table 5.1) are relatively wide to E_w of DExp, the total integrand is the distribution function, but as time increases it is progressively truncated from the left side by the sigmoid-like DExp. The location of maximum of the total integrand can significantly shifts toward the left side, and the shape of integrand gets skewed. On the other hand, the relatively narrow form of the continuous distribution function to the width of DExp makes the shape of total integrand like the distribution function, but the amplitude is drastically reduced by DExp as time proceeds. The nature (symmetrical or asymmetrical) of the distribution function remains unchanged; however, the global maxima of the whole integrands get shifted. Conversely, pyrolysis is not solely decomposition of the biomass, but the simultaneous degradation of various constituents (moisture content, hemicellulose, lignin, and cellulose). Therefore, it is difficult to describe pyrolysis through a single distribution function. It can also be an alternative to combine the two limits together for two distinct regions of experimental curve. But it is more important to look at the behavior of thermoanalytical data sets rather than applying "hit–and-trial" scheme. The data set is highly unpredictable during devolatization stage [29] and the reason is some fast-autocatalytic reactions that trigger the secondary pyrolysis, which is difficult to model through the conventional form without any transmutation, or variation of limit on σ (standard deviation), as the dependency of such reactions relies on the residence time of volatiles [25] inside the cellulose matrix; however, there are also some other factors, such as grain orientation, size of biomass particle, structural effect, and moisture content. Thus, the model must be robust enough to explain the transition states that depend on the residence time of volatile. Although the release of volatile content changes if heating rate varies from ramping to constant temperature cases, so there is no appreciable variation encountered in the amount of released volatile until the two parts of the DAEM integral overlap completely [3].

DExp as well as distribution together must be varied with time and temperature. Analysis is carried out for the relatively narrow distribution function ($\sigma \to 0$). The development in the subsequent sections pivoted around the method of the moving maximum for the Laplace integral problems (Refer Appendices). The behavior of the density function as well as DExp is examined at the interior points of activation energy domain ($p''(E) > 0$). The objective of implementing technique is to determine the moving maxima of the whole integrand as a function of time so that the TG data can be interpreted easily.

5.2.2.1 Asymptotic scheme

The postulation for the narrow distribution of the continuous distribution of types (Table 5.1) is taking standard deviation $\sigma = RT_1$, where T_1 denotes temperature at the inflexion point and follows the given thermal history. The local maxima y^e of corresponding functions ($M(y)$, $J(y)$, $U(y)$, and $H(y)$) of distribution functions is sought by finding the point where $M'(y)$, $J'(y)$, $U'(y)$, and $H'(y)$ are zeros. Asymptotes and shapes of transmuted function are decided by following the two limits [49].

Proposition 5.1. *The asymptotic of eq. (5.1) if $p(E) \to 0$ are*

$$\text{(i) } \boldsymbol{p^t(E) \sim (1+\tau)p(E)}$$

$$\text{(ii) } \boldsymbol{G^t(E) \sim (1+\tau)G(E)}$$

Proposition 5.2. *The asymptotic of eq. (5.1) as $E \to \infty$ are*

$$\text{(i) } \boldsymbol{p^t(E) \sim (1-\tau)p(E)}$$

$$\text{(ii) } \boldsymbol{G^t(E) \sim (1-\tau)G(E)}$$

Modeling and analyzing material properties in the field of engineering are very complicating and cumbersome. Sometimes, the data is too big to predict precisely through the conventional distribution function, especially the life prediction of machine parts. The density functions are widely implemented for studying fatigue and endurance life of engineering device and materials. Despite the wide range of application of distribution function, sometimes it is unable to explain different types of data that has nonmonotonic attribute; for instance, conversion of material at different thermal histories. Therefore, it is required to develop a flexible parametric model that can appropriately be embedded into a large model by adding an additional shape parameter. A significant contribution has been carried out for transmutation of the conventional model, thus enhancing flexibility in modeling of various types of life prediction data of the engineering components. The new transmutation models are developed and comprehensively perused in some of literatures [35, 36, 38, 49]. Therefore, some of the distribution types are modified to predict the thermoanalytical data efficiently through activation energy modeling.

5.2.2.1.1 Case A: transmuted Lindley (refer Appendix A)

Invoking eq. (5.1) and replacing the distribution function with transmuted function $p^t(y_1)$ by proper transformation in eq. (5.2), we have

$$1-\alpha \sim \int_0^\infty \gamma \exp\left\{-\exp\left(\frac{y_s-y_1}{y_{w1}}\right)\right\} p^t(y_1)dy_1 \tag{5.10}$$

Putting the value of $p^t(y_1)$, we get

$$1-\alpha \sim \gamma\left[\int_0^\infty \exp\left\{-\exp\left(\frac{y_{s1}-y_1}{y_{w1}}\right)\right\}(kf_1(y_1)+(1-k)f_2(y_1))dy_1\right] \tag{5.11}$$

Equation (5.11) can be written as a convex combination of functions $f_1(y_1)$ and $f_2(y_1)$:

$$\begin{aligned}1-\alpha \sim \Bigg[\int_0^\infty \exp\left\{-\exp\left(\frac{y_s-y_1}{y_w}\right)\right\}\Bigg(k\frac{\omega_1{}^2}{\theta}\psi(y_1-1)\exp\{-\omega_1(y_1-1)\}\\ +(1-k)\omega_1\theta\exp\{-\omega_1(y-1)\}\Bigg)dy_1\Bigg]\end{aligned} \tag{5.12}$$

where k is a mixing fraction.

Rewrite eq. (5.12) into two different parts and solve it for the same value of y_1:

$$1-\alpha = \{(1-\alpha_0)+(1-\alpha_1)\} \tag{5.13}$$

Solve eq. (5.13) separately for $1-\alpha$.

Let,

$$1-\alpha_0 = \int_0^\infty \frac{k\omega_1{}^2}{\theta}\exp\left\{-\exp\left(\frac{y_s-y_1}{y_w}\right)\right\}\psi(y_1-1)\exp\{-\omega_1(y_1-1)\}dy_1 \tag{5.14}$$

Suppose that

$$V(y_1) = -\exp\left(\frac{y_s - y_1}{y_w}\right) - \omega_1(y_1 - 1) \tag{5.15}$$

The function $V(y_1)$ is at its extremum if $V'(y_1) = 0$.

Differentiating eq. (5.15) w.r.t y_1, we get

$$V'(y_1{}^e) = \frac{\exp\left(\frac{y_{s1} - y_1{}^e}{y_{w1}}\right)}{y_{w1}} - \omega_1$$

Let the location of maximum is denoted by $y_1{}^e$, then

$$y_1{}^e = y_{s1} - y_{w1}\ln(\omega_1 y_{w1})$$

The second derivative of eq. (5.15) w.r.t. y_1 is given by

$$|V''(y_1{}^e)| = \frac{\omega_1}{y_{w1}}$$

Replacing $V(y_1)$ in eq. (5.15) by its Taylor expansion, we have

$$V(y_1) \sim V(y_1{}^e) + \frac{(y - y_1{}^e)^2}{2} V''(y_1{}^e) + \cdots$$

After simplification, we get

$$(1-\alpha_0) \sim 1.25(y_1{}^e - 1)k\left(\psi\frac{\omega_1}{\theta}\sqrt{y_{w1}\omega_1}\right)\left(\operatorname{erf}\left(y_1{}^e\sqrt{\frac{0.5\omega_1}{y_{w1}}}\right) + 1\right)\exp\{-\omega_1(y_{w1} + y_1{}^e - 1)\} \tag{5.16}$$

Now consider the other term $(1 - \alpha_1)$

$$1 - \alpha_1 = \int_0^\infty \theta\omega_1(1-k)\exp\left\{-\exp\left(\frac{y_{s1} - y}{y_{w1}}\right)\right\}\theta\exp\{-\omega_1(y-1)\}dy \tag{5.17}$$

Similarly, the required expression for $(1 - \alpha_1)$ is,

$$(1-\alpha_1) \sim 1.25\,\theta(1-k)\left(\sqrt{y_{w1}\omega_1}\right)\left(\operatorname{erf}\left(y_1{}^e\sqrt{\frac{0.5\omega_1}{y_{w1}}}\right) + 1\right)\exp\{-\omega_1(y_{w1} + y_1{}^e - 1)\} \tag{5.18}$$

Substituting eqs. (5.16) and (5.17) into eq. (5.13), we have the resultant solution for transmuted Lindley function at different limits

$$\lim_{p(E)\to 0}(1-\alpha\ ){\sim}1.25\ (1+\tau)(1-k)\sqrt{y_{w1}\omega_1}\left(\operatorname{erf}\left(y_1{}^e\sqrt{\frac{0.5\omega_1}{y_{w1}}}\right)+1\right)$$

$$\exp\{-\omega_1(y_{w1}+y_1{}^e-1)\}\{\theta+\omega_1(y_1{}^e-1)\}$$

$$\lim_{E\to\infty}(1-\alpha\ ){\sim}1.25\ (1-\tau)\ (1-k)\sqrt{y_{w1}\omega_1}\left(\operatorname{erf}\left(y_1{}^e\sqrt{\frac{0.5\omega_1}{y_{w1}}}\right)+1\right)$$

$$\exp\{-\omega_1(y_{w1}+y_1{}^e-1)\}\{\theta+\omega_1(y_1{}^e-1)\}$$

5.2.2.1.2 Case B: transmuted Weibull (refer Appendix B)

From eqs. (5.1) and (5.2), we get

$$1-\alpha=\int_0^\infty y\exp\left\{-\exp\left(\frac{y_s-y}{y_{w2}}\right)\right\}p^t(y_2)dy \tag{5.19}$$

Substituting the density ($p(y_2)$ from Table 5.1 into eq. (5.1) for the Weibull distribution, we have

$$1-\alpha=(1+\tau)\omega_2{}^\beta\beta\int_0^\infty(y_2-1)^{\beta-1}\exp\left\{-\exp\left(\frac{y_{s2}-y_2}{y_w2}\right)-\left((\omega_2(y_2-1))^\beta\right)\right\}dy \tag{5.20}$$

The final expression after transmutation is expressed through eq. (5.20).

Assume that $M(y_2)$ is a function whose maxima is at $y_2=y_2{}^e$, then

$$M(y_2)=-\exp\left(\frac{y_{s2}-y_2}{y_{w2}}\right)-\left((\omega_2(y_2-1))^\beta\right) \tag{5.21}$$

Find the first- and the second-order derivatives of eq. (5.21) and substitute its Taylor series expansion into eq. (5.20):

$$M'(y_2)=\frac{\exp\left(\frac{y_s-y_2}{y_w}\right)}{y_w}-\beta\omega_2{}^\beta\left(y_2-1\right)^{\beta-1}$$

Equating $M'(y_2)=0$, we get

$$y_2{}^e=1+(\beta-1)y_{w2}W\left(\frac{1}{(\beta-1)y_{w2}}\left(\frac{1}{\beta y_{w2}\omega_2{}^\beta}\right)^{\frac{1}{(\beta-1)}}\exp\left\{\frac{1}{(\beta-1)}\left(\frac{y_{s2}-1}{y_{w2}}\right)\right\}\right)$$

Likewise,

$$M''(y_2)=-\frac{\exp\left(\frac{y_{s2}-y_2}{y_{w2}}\right)}{y_{w2}{}^2}-\beta(\beta-1)\omega_2{}^\beta(y_2-1)^{\beta-2} \tag{5.22}$$

$$M(y_2) \sim M(y_2{}^e) + \frac{(y - y_2{}^e)^2}{2} M''(y_2{}^e) + \cdots \tag{5.23}$$

After substituting $M''(y_2)$ from eq. (5.22) into eq. (5.23) and replacing $M(y_2)$ by its expansion in eq. (5.20), we get the asymptotic expression for the Weibull distribution for different limits:

$$\lim_{p(E)\to 0} (1-\alpha) \sim (1+\tau) \left(0.707 \sqrt{\frac{\pi \omega_2{}^\beta \beta (y_2{}^e - 1)^\beta}{\left\{ \left(\frac{y_2{}^e - 1}{y_{w2}} \right) + (\beta - 1) \right\}}} \right) (\mathrm{erf}(y_2{}^e) + 1)$$
$$\exp\left\{ -\omega_2{}^\beta (y_2{}^e - 1)^{\beta - 1} (y_2{}^e + \beta y_{w2} - 1) \right\}$$

$$\lim_{E\to\infty} (1-\alpha) \sim (1-\tau) \left(0.707 \sqrt{\frac{\pi \omega_2{}^\beta \beta (y_2{}^e - 1)^\beta}{\left\{ \left(\frac{y_2{}^e - 1}{y_{w2}} \right) + (\beta - 1) \right\}}} \right) (\mathrm{erf}(y_2{}^e) + 1)$$
$$\exp\left\{ -\omega_2{}^\beta (y_2{}^e - 1)^{\beta - 1} (y_2{}^e + \beta y_{w2} - 1) \right\}$$

5.2.2.1.3 Case C: Transmuted Gaussian (refer Appendix C)

Replacing the Gaussian distribution from eq. (5.2) with transmuted Gaussian function, we have

$$1 - \alpha = \int_0^\infty E_0 \exp\left\{ -\exp\left(\frac{y_s - y_3}{y_w} \right) \right\} p^t(y_3) dy_3 \tag{5.24}$$

After putting the value of the Gaussian distribution from Table 5.1 into eq. (5.1) and substituting into eq. (5.24), the transmuted form will be represented as

$$1 - \alpha = (1+\tau) \sqrt{\frac{\omega_3}{\pi}} \int_0^\infty \exp\left\{ -\exp\left(\frac{y_s - y_3}{y_w} \right) - \omega_3 (y_3 - 1)^2 \right\} dy_3 \tag{5.25}$$

Suppose that

$$J(y_3) = -\exp\left(\frac{y_s - y}{y_w} \right) - \omega_3 (y_3 - 1)^2$$

The first and the second derivatives of $J(y_3)$ w.r.t "y_3" are given by eqs. (5.26) and (5.27), respectively

$$J'(y_3{}^e) = \frac{\exp\left(\frac{y_s - y}{y_w} \right)}{y_w} - 2\omega_3 (y_3 - 1) \tag{5.26}$$

$$|J''(y)| = 2\omega_3\left(\frac{{y_3}^e - 1}{y_w} + 1\right) \tag{5.27}$$

The location of maximum of function $J(y_3)$ is obtained by $J'(y_3) = 0$. Hence

$${y_3}^e = 1 + y_w W\left(\frac{1}{2\omega_3 {y_w}^2}\exp\left(\frac{y_s - 1}{y_w}\right)\right)$$

Substituting the values of $|J''({y_3}^e)|$ in the Taylor series expansion of $J(y_3)$ about ${y_3}^e$, we have

$$J(y_3) \sim J({y_3}^e) + \frac{(y - {y_3}^e)^2}{2} J''({y_3}^e) + \cdots$$

After simplification and rearrangement, the required solution of transmuted Gaussian function is expressed as

$$\lim_{p(E)\to 0}(1-\alpha) \sim (1+\tau)(\mathrm{erf}({y_3}^e)+1)\left\{\frac{\exp\{-\omega_3({y_3}^e-1)(2y_w+{y_3}^e-1)\}}{\sqrt{\left(\frac{{y_3}^e-1}{y_w}+1\right)}}\right\} \text{ for } p(E) \to 0$$

$$\lim_{E\to\infty}(1-\alpha) \sim (1-\tau)(\mathrm{erf}({y_3}^e)+1)\left\{\frac{\exp(\{-\omega_3({y_3}^e-1)(2y_w+{y_3}^e-1)\})}{\sqrt{\left(\frac{{y_3}^e-1}{y_w}+1\right)}}\right\} \text{ for } E \to \infty$$

5.2.2.1.4 Case D: Transmuted Frechet (refer Appendix D)

The expression of transmuted Frechet is obtained after replacing density function from transmuted density function.

Equation (5.28) is the required expression which is subjected to investigation:

$$1 - \alpha \sim \int_0^\infty \mu \exp\left\{-\exp\left(\frac{y_{s3} - y_4}{y_{w3}}\right)\right\} p^t(y_4)\,dy_4 \tag{5.28}$$

Replacing transmuted expression through substituting the value of density function from Table 5.1, we have

$$1 - \alpha \sim (1 + |\tau|)\int_0^\infty \exp\left[\left\{-\exp\left(\frac{y_{s3} - y}{y_{w3}}\right)\right\}\rho {\omega_4}^{-\rho}(y_4 - 1)^{-(1+\rho)} e^{-\{\omega_4(y_4-1)\}^{-\rho}}\right] dy_4 \tag{5.29}$$

To solve eq. (5.29), suppose that

$$U(y_4) = -\exp\left(\frac{y_s - y_4}{y_{w3}}\right) - \{\omega_4(y_4 - 1)\}^{|\rho|} \tag{5.30}$$

$$1 - \alpha \sim (1 + |\tau|) \int_0^{\infty} \exp(U(y_4)) dy_4 \tag{5.31}$$

Obtain the first- and the second-order derivatives of $U(y_4)$ as it is performed in the other cases:

$$U'(y_4{}^e) = \frac{\exp\left(\frac{y_s - y_4{}^e}{y_{w3}}\right)}{y_{w3}} - |\rho|\omega_4{}^{|\rho|}(y_4{}^e - 1)^{|\rho - 1|}$$

The location of maximum value is derived by $U'(y_4) = 0$.

$$y_4{}^e = 1 + y_{w3} W\left(\frac{1}{|\rho - 1| y_{w3}} \left(\frac{1}{y_{w3}|\rho|\omega_4{}^{|\rho|}}\right)^{\frac{1}{|\rho - 1|}} \exp\left(\frac{y_s - 1}{y_{w3}}\right)\right)$$

After simplifying the second-order derivative of eq. (5.30), we get

$$\left|U''(y_4{}^e)\right| = |\rho|(y_4{}^e - 1)^{|\rho - 2|}\omega_4{}^{|\rho|}\left(\frac{(y_4{}^e - 1)^{|\rho - 1|}}{y_{w3}} + |\rho - 1|\right)$$

Expand $U(y_4)$ with the Taylor series expression about $y_4 = y_4{}^e$,

$$U(y_4) \sim U(y_4{}^e) + \frac{(y - y_4{}^e)^2}{2} U''(y_4{}^e) + \cdots \tag{5.32}$$

After replacing $U(y_4)$ from its Taylor expansion given by eq. (5.32), we get the solution for the transmuted Frechet as

$$\lim_{p(E)\to 0} (1 - \alpha) \sim (1 + \tau) 0.707 \sqrt{\frac{\pi (y_4{}^e - 1)^{|\rho|}\omega_4{}^{|\rho|}}{\left(\frac{(y_4{}^e - 1)^{|\rho - 1|}}{y_{w3}} + |\rho - 1|\right)}} (\mathrm{erf}(y_4{}^e) + 1)$$

$$\left[\exp\left\{-\omega_4{}^{|\rho|}(y_4{}^e - 1)^{|\rho - 1|}(y_{w3}|\rho| - (y_4{}^e - 1))\right\}\right]$$

$$\lim_{E\to\infty} (1 - \alpha) \sim (1 - \tau) 0.707 (\mathrm{erf}(y_4{}^e) + 1) \sqrt{\frac{|\rho|\pi (y_4{}^e - 1)^{|\rho|}}{\left(\frac{(y_4{}^e - 1)^{|\rho - 1|}}{y_{w3}} + |\rho - 1|\right)}}$$

$$\left[\exp\left\{-\omega_4{}^{|\rho|}(y_4{}^e - 1)^{|\rho - 1|}(y_{w3}|\rho| - (y_4{}^e - 1))\right\}\right]$$

Note: Valid for the negative shape parameter ($\rho < 0$).

5.2.2.1.5 Case E: Transmuted Rayleigh distribution (refer Appendix E)

Two-parameter Rayleigh distribution is used for demonstrating the scheme. The transmuted distribution function is obtained by replacing the density function from its expression given in Table 5.1. Equation (5.33) is the transmuted expression for first-order chemical reactions:

$$1-\alpha \sim \int_0^\infty \phi \exp\left\{-\exp\left(\frac{y_{s4}-y}{y_{w4}}\right)\right\} p^t(y_5)dy \tag{5.33}$$

After obtaining $p^t(y_5)$ from eq. (5.1), we get the expression for the expected solution

$$1-\alpha \sim \int_0^\infty \exp\left\{-\exp\left(\frac{y_{s4}-y}{y_{w4}}\right)\right\} 2\omega_5(y-1)\exp\left\{-\omega_5(y_5-1)^2\right\}dy_5 \tag{5.34}$$

Assume that

$$H(y_5) = -\exp\left(\frac{y_{s4}-y}{y_{w4}}\right) - \omega_5(y_5-1)^2$$

Then eq. (5.34) can be represented as

$$1-\alpha \sim \int_0^\infty 2\omega_5\phi(y-1)\exp(H(y_5))dy_5 \tag{5.35}$$

The first derivative of $H(y_5)$ w.r.t. "y_5" is

$$H'(y_5{}^e) = \frac{\exp\left(\frac{y_{s4}-y}{y_{w4}}\right)}{y_{w4}} - 2\omega_5(y_5{}^e-1)$$

For deriving the location of maximum, equate $H'(y_5) = 0$

$$y_5{}^e = 1 + y_{w4} W\left(\frac{1}{2\omega_5 y_{w4}{}^2}\exp\left(\frac{y_{s4}-1}{y_{w4}}\right)\right)$$

"y_5e" is the location where the function $H(y_5)$ attains its maximum.

Similarly, after simplification, the second derivation is obtained as

$$H''(y_5{}^e) = \left|2\omega_5\left(\frac{(y_5{}^e-1)}{y_{w4}}+1\right)\right| \tag{5.36}$$

The Taylor series expansion of function $H(y_5)$ about $y_5{}^e$ is given by

$$H(y_5) \sim H(y_5{}^e) + \frac{(y-y_5{}^e)^2}{2} H''(y_5{}^e) + \cdots \tag{5.37}$$

After putting the value of $H''(y_5{}^e)$ from eq. (5.36) into eq. (5.37) and replacing $H(y_5)$ in eq. (5.35) with its Taylor series expansion (eq. (5.37)), we have the required expression for the transmuted Rayleigh distribution:

$$\lim_{p(E)\to 0} 1-\alpha \sim (1+\tau)\left(\sqrt{\frac{\omega_{5\pi}}{2\left(\frac{(y_5{}^e-1)}{y_{w4}}+1\right)}}\right)(\mathrm{erf}(y_5{}^e)+1)\exp\{-\omega_5(y_5{}^e-1)\ (2y_{w4}+y_5{}^e-1)\}$$

$$\lim_{E\to\infty} 1-\alpha \sim (1-\tau)\left(\sqrt{\frac{\pi\omega_5}{2\left(\frac{(y_5{}^e-1)}{y_{w4}}+1\right)}}\right)(\mathrm{erf}(y_5{}^e)+1)\exp\{-\omega_5(y_5{}^e-1)\ (2y_{w4}+y_5{}^e-1)\}$$

5.3 Application and computation analysis of TG data

Elemental analysis of sample subjected to thermal degradation is conducted through CHNS-O analyzer. The CHNS-O analyzer (vario MACRO) is used to evaluate the chemical composition of pine needles. Before using the CHNS-O analyzer, the furnace is heated up to 1,473 K for 30 min. Once it gets heated, the capsule forms of sample are fed into a slot of rotating disk. The flow of oxygen maintains the catalytic combustion, whereas helium gas acts as a carrier gas stream so that the volatile gases with water (CO_2, SO_2, NO_2, and H_2O) can be absorbed at different reduction columns. These tubes are in-between combustion chamber and detection unit (computer-based software), where the gas mixture is separated into its components through purge or trap chromatography. Thereafter, each component is separately detected by thermal conductivity detector (TCD). The gases CO_2, H_2O, and SO_2 are absorbed in a sequence, whereas N_2 passes through all the three columns and it is finally detected by TCD. TCD generates the electrical signal that is proportional to the concentration of elementary components of biomass. The reference material is considered to be the powder of birch leaf. The bomb calorimeter at constant volume is used to measure the calorific value of pine needles. Thermal evaluation of pine waste is based on TG differential thermal analyzer (Diamond TG/DTA, Perkin Elmer, USA). The sample of 5.68 mg is placed in an electrobalance from 303 to 923 K. To avoid the measurement error due to buoyant effect at the high temperature, the horizontal differential-type balance is considered for experimentation. Thermocouple-type "R" is referred to measure the temperature inside the furnace. The volumetric flow of noninteractive gas is set at 200 mL min^{-1}. Indium and tin are used as reference materials for differential thermograms.

Table 5.2 illustrates the elemental characteristics and higher heating value of pine waste on the dry basis. TG data of pine sample is used to implement the transmutation of DAEM, so that qualitative assessment of different distribution functions can be performed. Computational as well as algorithm designing is carried out on MATLAB 2017b. The nonlinear thermal history is tested for narrow distribution nature of density functions. The different assigned values of parameters are repeatedly

Table 5.2: Elemental composition of pine needles.

C%	H%	N%	O%	S%	Ash content (%)	Higher heating value (MJ/kg)
53.7	6.48	0.56	36.13	0.10	2.92	22.23

examined in every iterative loop until it does not satisfy the given condition. If any parametric value does not validate the constraint, system searches for different values. Once the solution is obtained, the program terminates itself.

5.4 Results and discussion

The numerical solution of transmuted modeling of DAEM is performed by using the Laplace integral method. The proposed scheme is exhaustively examined for different distribution functions. The significant role of moving maximum with time on the energy distribution pattern of released volatile is illustrated graphically. There are three different limits imposed on each distribution case and the results are derived separately after transmutation.

5.4.1 Effect of transmutation on the Lindley distribution

The effect of transmuted Lindley distribution is shown in Figure 5.1(a)–(f). The sensitively of the Lindley distribution with respect to time and temperature is relatively higher than that of other types of distribution functions analyzed for the energy modeling purpose. There is a steep increase in the conversion rate of biomass with variation of kinetic parameters in Figure 5.1(a)–(f). Elevation of frequency factor causes the remaining mass proportion curves to shift upward with relatively high (Table A.1) residual mass of biomass. However, the variation of residual mass of biomass is controlled to some extent through transmutation, but relative change in the residual mass of biomass is negligibly small. Transmutation of the Lindley distribution infers that the higher activation energy of 103 kJ/mol can be easily modeled through Lindley distribution. With increase in activation energy, the remaining mass fraction gets shifted down; however, there is no significant change in the attribute of $(1-\alpha)$ curves with the variation in the thermal history of sample. The maximum upper limit of activation should not be more than $E_\infty \leq 5.72 \times 10^3$ kJ/mol, while the average activation energy should not be more than 103. Comparative sketch of the transmuted Lindley distribution with the classical DAEM is shown in Figure 5.2. The fact concluded through the Lindley distribution is that Lindley distribution is not able to describe the pyrolysis mechanism more effectively than the other distribution types, despite having the

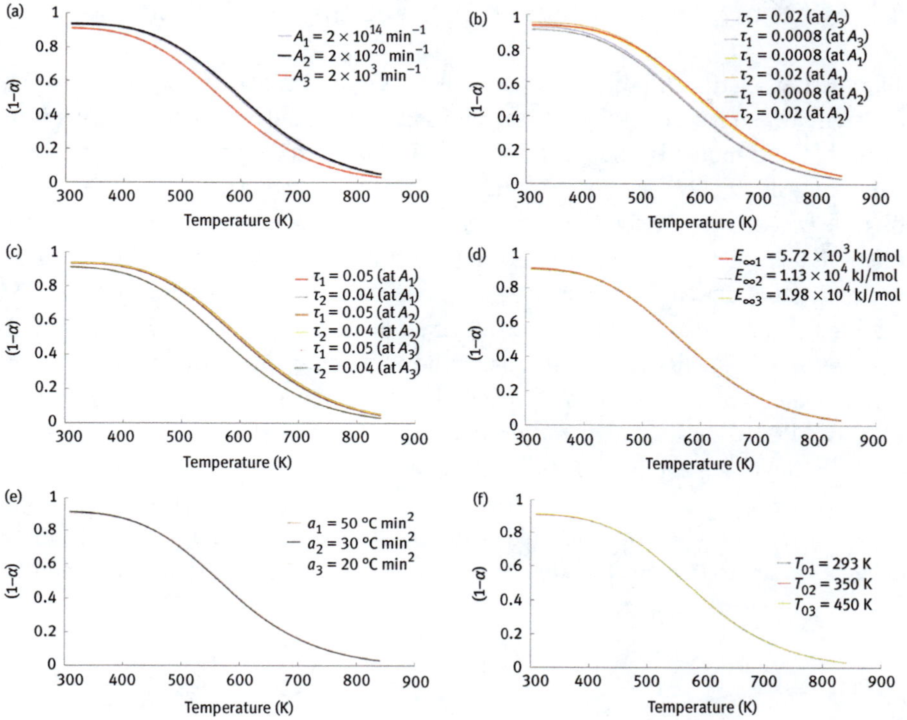

Figure 5.1: The effect of transmutation on the numerical solution of Lindley-based DAEM: (a)–(d) kinetic parameters; e and (f) characteristic of nonlinear thermal profile.

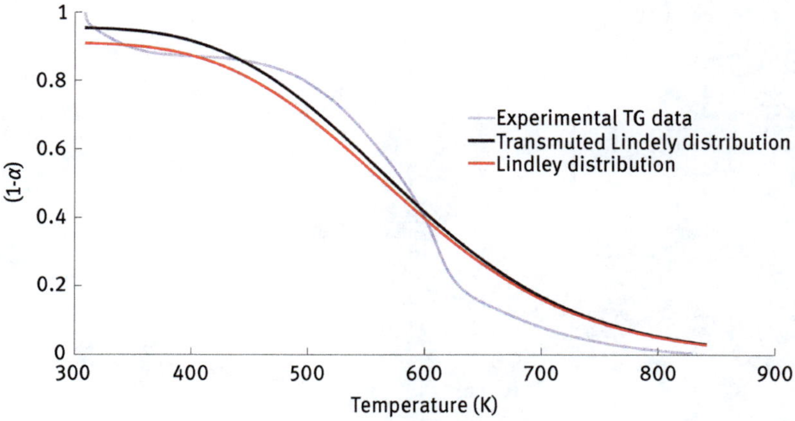

Figure 5.2: Comparative sketch of t-DAEM (Lindley) with the conventional DAEM form.

highest value of coefficient of regression. It may be possible that the Lindley distribution is not able to extrapolate pyrolysis effectively, and behavior is like the phenomenological models.

5.4.2 Effect of transmutation on the Weibull distribution

Unlike the Gaussian distribution, the Weibull distribution is abounded with different parameters, which is also one of the reasons of higher redundancy than the Gaussian. The effect of transmuted Weibull distribution is illustrated in Figure 5.3(a)–(f). The thermal decomposition of biomass takes some time to reach the critical temperature for initiating the process. The Weibull distribution is unable to distinguish the response time at different activation energies; therefore, all the curves are highly overlapped without any significant variation at the end states of pyrolysis (Figure 5.3(a)). The effect of different limit on the transmuted (Table B.1) Weibull distribution is depicted in Figure 5.3 (b)–(f). As the frequency factor increases, the remaining mass fraction curves get shifted upward, which results in shifting of critical decomposition temperature. Unlike the transmuted Gaussian distribution, the curves are converging for the lower range of activation energies, or the relative width between activation energies of parallel reaction must be narrow for the better simulation of pyrolysis through Weibull distribution. At the elevated heating rate with time, the devolatilization stage gets shifted to the left. The similar trend is obtained for the initial experimental

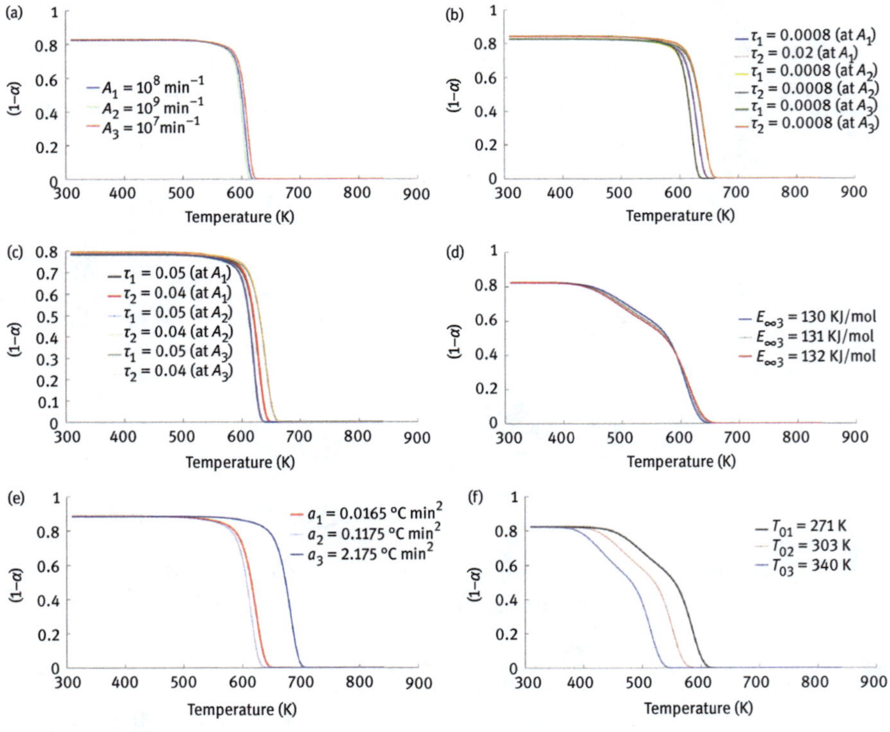

Figure 5.3: The effect of transmutation on the numerical solution of the Weibull-based DAEM: (a)–(d) kinetic parameters; (e) and (f) characteristic of nonlinear thermal profile.

temperature. Although the Weibull distribution is not able to interpolate the char formation stage effectively, despite the devolatilization stage effectively described the transmuted Weibull distribution. But the objective must not be overfitting of simulated solution to the thermoanalytical data, as unfamiliarity with the mechanism of autocatalytic reactions and the residence time of volatiles may provide erroneous solution. The maximum upper limit of activation energy for the transmuted Weibull distribution must not be more than 130 kJ/mol and the average activation energy (kJ/mol) should be within the domain of (69, 89] for the better convergent rate. Comparative demonstration of two different schemes is shown in Figure 5.4. The transmuted form of Weibull describes the devolatization stage of decomposition of biomass efficiently, but the residence time of volatile content increases with temperature and the whole biomass is transformed into volatile gases (Figure 5.4). The char yield became asymptotic with temperature.

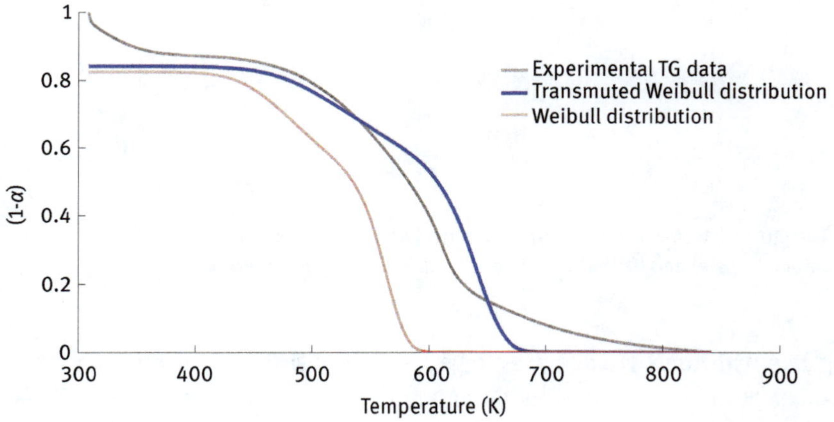

Figure 5.4: Comparative sketch of t-DAEM (Weibull) with the conventional DAEM form.

5.4.3 Effect of transmutation on the Gaussian distribution

The comprehensive effect of transmutation on the Gaussian DAEM is depicted graphically in Figure 5.5(a)–(f). At the initial stage of pyrolysis, the remaining mass fraction ($1-\alpha$) of biomass is always near to 1, unless there is interactive atmosphere or adsorption, or surface oxidation processes occur during thermal degradation of biomass. But the graphical representation of the classical DAEM (Figure 5.5(a)) is exactly 1, which is quite idealistic for pragmatic pyrolysis condition. As the frequency factor increases, there is an appreciable displacement of the upper segments of remaining mass fraction ($1-\alpha$) than that of lower halves at the constant rate of change

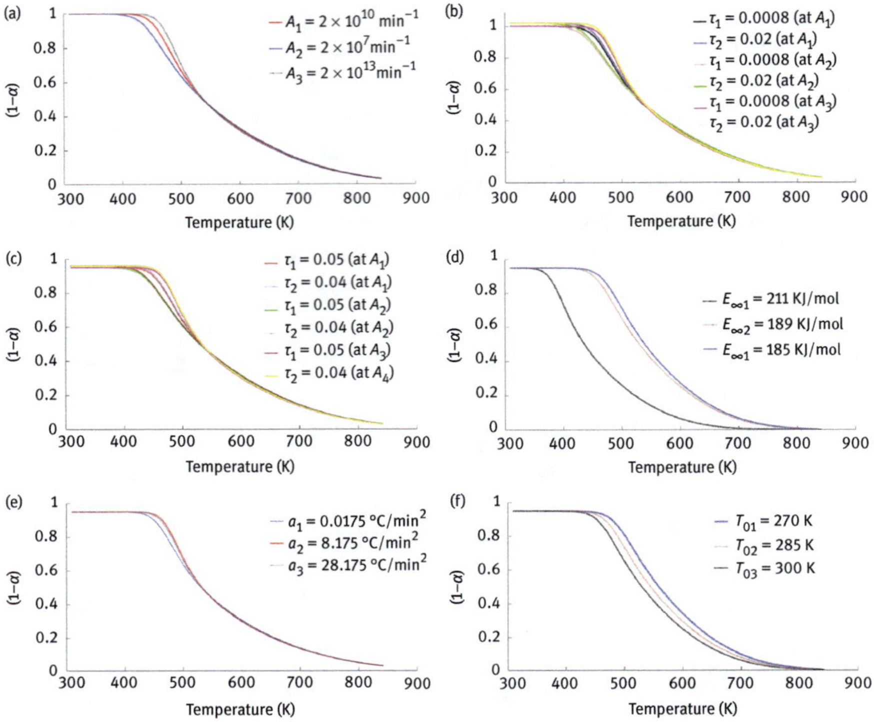

Figure 5.5: The effect of transmutation on the numerical solution of the Gaussian-based DAEM: (a)–(d) kinetic parameters; (e) and (f) characteristic of nonlinear thermal profile.

of heating rate with time. Despite drastic variation of distribution parameters, converging ability of the numerical solution is improved through transmutation. Thus, the homogeneity of shape is maintained throughout the process. The most important facet of transmutation is illustrated in Figure 5.5(b) and (c). The qualitative aspect of transmutation of distribution of activation energy is known through the characteristic of explanatory variable rather than the coefficient of regression analysis. The overlapping of curves happens if the density function of activation energy approaches zero, whereas the nonconvergent solution converges to solution, if activation energies spectrum is very broad for the Gaussian distribution; therefore, the Gaussian distribution can be the best fitting model for those reactions (Table C.1) where the activation energy variation among reactions is colossal. The effect of initial temperature gets devolatization stage shifted to the right without any significant variation in "drying" as well as "char formation" stages. The residual mass obtained after simulating the pyrolysis process is approximately the same with $\pm 1.8\%$ difference. The effect of activation energy variation is depicted in Figure 5.5(d). The increase in the activation energy causes the intermediate stages to shift left. According to the analysis, the maximum upper bound of activation energy should not be more than $E_\infty \leq 185$ kJ/mol and the maximum permissible limit for the average value of activation energy (kJ/mol) must lie in the domain of [120, 150). Inclusion of additional parameters supports the convergence rate of the numerical solution obtained by the Gaussian distribution. The effect of transmutation is mainly on the intermediate states of decomposition, which plays a pivotal role in the kinetic mechanism of pyrolysis. However, the influence of transmutation on the end stages of pyrolysis is difficult to interpolate, yet the variation gap and derivation of the classical DAEM from practical aspect of pyrolysis are overshadowed by transmutation and the form becomes more homogeneous than the classical form of DAEM. The nature of Arrhenius parameters and its effect on the thermal decomposition of biomass can be easily distinguished through transmutation. Comparative analysis of transmuted DAEM with classical DAEM is depicted in Figure 5.6. It clearly demarcated the controlling ability of transmutation to describe pyrolysis process far more effectively than the classical DAEM.

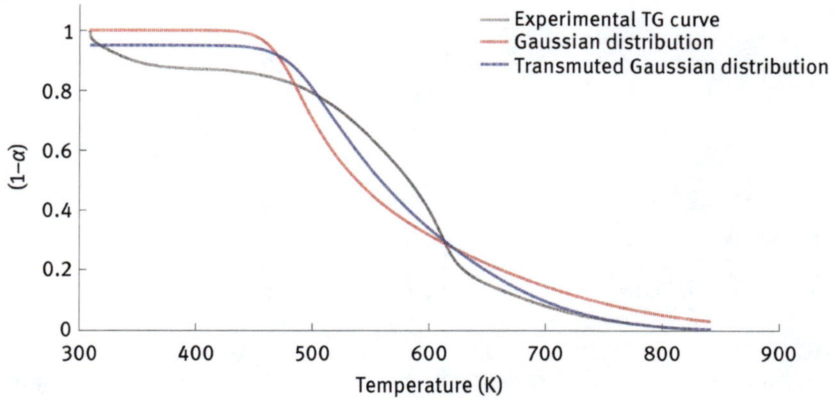

Figure 5.6: Comparative sketch of t-DAEM (Gaussian) with the conventional DAEM form.

5.4.4 Effect of transmutation on the Frechet distribution

Graphical representation of the transmuted Frechet distribution is illustrated in Figure 5.7 (a)–(f). The increase in the frequency factor increases the conversion rate, which results in residual mass approach to the experimental condition at the reduced timescale. However, the process is accelerated through introduction of linear mixing, and the remaining mass curves cause to shift upward as $E \to 0$. Therefore, the transmuted Frechet holds validation for set of reactions having relatively low variation of activation energies. The upper limit of activation energy is 190 kJ/mol, while the average activation energy (E kJ/mol) must not be more than $E_0 \leq 120$. Activation energy modeling through transmuted Frechet distribution provides good result only at lower initial temperature, and the variation of heating rate must not be very high. The

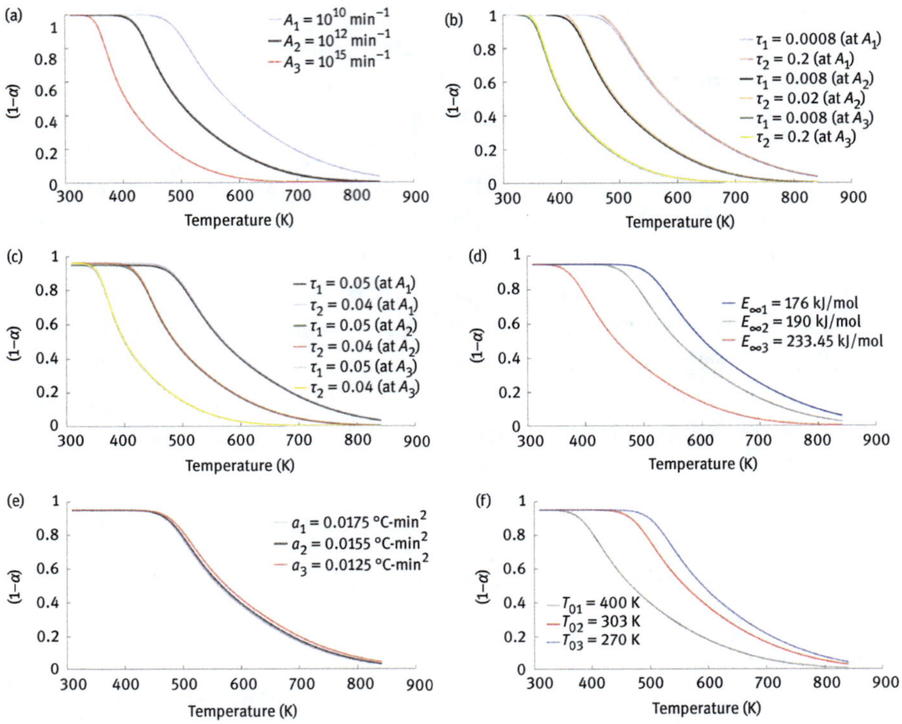

Figure 5.7: The effect of transmutation on the numerical solution of Frechet-based DAEM: (a)–(d) kinetic parameters; (e) and (f) characteristic of nonlinear thermal profile.

predicted transmuted DAEM solution with respect to the classical DAEM is shown in Figure 5.8. The transmuted Frechet distribution does not provide much better convergence than that of the Gaussian, the Weibull, and the Rayleigh distributions.

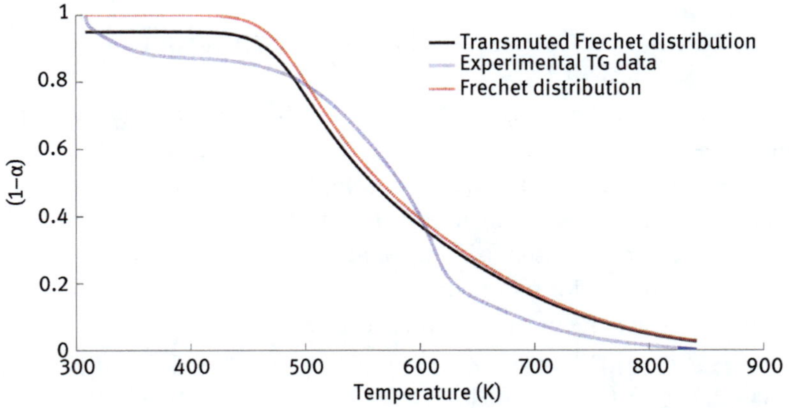

Figure 5.8: Comparative sketch of t-DAEM (Frechet) with the conventional DAEM form.

Comparative analysis of Laplace temperature with other numerical techniques is based on the attribute of numerical solution. However, the Copula distribution [1, 23] and 2-DAEM [5] are alternative forms of linear mixing of two different functions, but the emphasis is laid upon fitting of data rather than behavior of kinetic parameters with respect to time and the thermal history. Dynamic nature of inflexion points in the wide distribution case [21], and neglecting the time factor in the kinetic modeling, may provide a better fitting but the measured values would be quite far away from the pragmatic solution. Another way of solving the double integral of DAEM is through Simpson methods [21]. But without knowing the error component while spacing the activation energies leads to the fallacious inference. The concept of analyzing the thermal data is based on the stability of local maxima of curves and removing the inflexion point $H''(y_5{}^e) \neq 0$ out of the numerical solution. The derived solutions for the transmuted distributions rely upon the attribute of y^e with time. In the beginning, the moving maxima remains constant with temperature, but later it increases linearly as time proceeds. The narrow distribution is not able to interpolate the initial decomposition phase, as the Lambert W function is almost dormant in the beginning stage of simulation, while the Lindley distribution is related to the nature of logarithms. Therefore, spontaneity of reactions modeled through Lindley reactions is quite different from other transmuted forms. This problem can be tackled if the similar experiment is conducted at the complex thermal history. The obtained values of kinetic parameter for pine needles are homogenous with the isoconventional methods without any compensatory effect [29], which is a positive benchmark of the Laplace integral. However, the applicability of scheme is limited to only those functions that have the nonzero second derivatives and exhibit nonmonotonicity with time or temperature. The effect of proposed scheme is based on the significance of the whole integrand about its central value, which is overlooked in the previous studies [2, 21].

5.4.5 Effect of transmutation on the Rayleigh distribution

The effect of the transmutation on the activation energy modeling through the Rayleigh distribution is depicted in Figure 5.9(a)–(f). The resolution capacity of Rayleigh distribution with respect to variation of parameters is far more than that of the Gaussian and the Weibull distributions. One can observe a fine demarcation of behavioral variation of $(1-\alpha)$ with the frequency factor as illustrated in Figure 5.5(a). The increasing value of frequency increases the dehydration temperature and the $(1-\alpha)$ curves are deviated to the right side without any variation in the char formation temperature. However, the deviation of the remaining mass fraction is controlled through transmutation of Rayleigh distribution (Figure 5.9(b)–(f)). Likewise, the Gaussian distribution and the Rayleigh distribution support the reactions that occur at higher activation energies. There is no significant variation in the intermediate stages of decomposition except the time of initiation of reactions that increases with

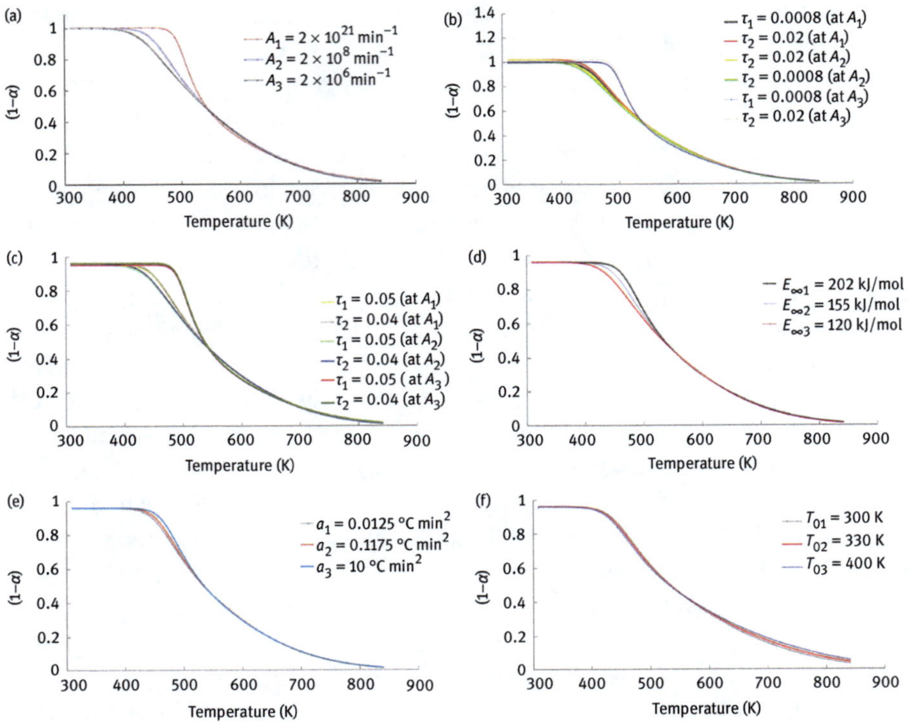

Figure 5.9: The effect of transmutation on the numerical solution of Rayleigh-based DAEM: (a)–(d) kinetic parameters; (e) and (f) characteristic of nonlinear thermal profile.

increasing activation energies, which is obviously much practical. The upper bound of activation energy variation for the effective representation of pyrolysis reaction is 220 kJ/mol, whereas the feasible (Table E.1) boundary limits of variation in average activation energy (kJ mol^{-1}) is within the range of (96, 127]. Increasing the initial experimental temperature causes the remaining mass fraction curves $(1–\alpha)$ to shift upward with higher residual mass than that of lower initial temperature. Similarly, the decomposition temperature at the initial stage of pyrolysis is elevated with drastic variation of heating rate with time (Figure 5.9(e)). However, perturbance of Rayleigh distribution due to ingression of linear mixing does not affect the form of distribution drastically (Figure 5.10).

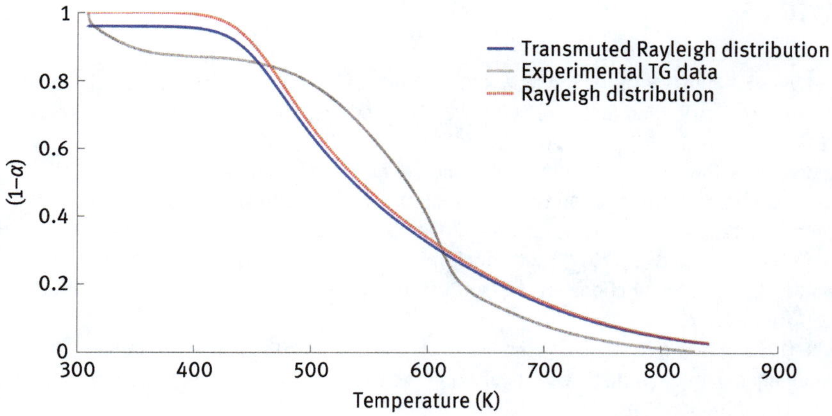

Figure 5.10: Comparative sketch of t-DAEM (Rayleigh) with the conventional DAEM form.

5.5 Conclusion

The narrow distribution of transmuted function is adopted to derive the numerical solution. Transmutation of the Gaussian distribution, the Weibull distribution, and the Rayleigh distribution provides the better results than that of transmutation form of Lindley and Frechet distributions. Every distribution function has its own unique attribute for representing the variation of activation energies. The Gaussian distribution and the Rayleigh distribution are found to be good for developing the model for higher activation energy reactions of maximum of 202 kJ/mol, while the Weibull distribution holds good relationship for the activation energy range of 80–130 kJ/mol. The modeling of activation energy through Frechet distribution is valid for the negative value of shape factor, whereas Lindley has anomality with the measured value of pine needles from the isoconversional method. However, there is a likelihood of existence of those reactions whose activation energy is of 103 kJ/mol. The maximum value of activation energy for the transmuted Frechet distribution is found to be 190 kJ/mol. The frequency factor for all the cases is estimated to lie in domain of 103–1,013 min^{-1}. The variation of activation energy with time, temperature, energy spacing, and thermal profile and the significant contribution of the whole integrand are investigated through knowing the attribute of moving maximum of the whole integrand. However, the computational effort of scheme is relatively high to other schemes [1, 23, 50], as algorithm evaluates the location of maximum with respect to time.

References

[1] Dhaundiyal A., Singh S.B., & Hanon M.M. (2018) Study of distributed activation energy model using bivariate distribution function, f (E 1, E 2), Thermal Science and Engineering Progress, 5, 388–404. doi: 10.1016/j.tsep.2018.01.009.

[2] Dhaundiyal A. & Tewari P. (2017) Kinetic parameters for the thermal decomposition of forest waste Using Distributed Activation Energy Model (DAEM), Environmental and Climate Technologies, 19, 15–32. doi: 10.1515/rtuect-2017-0002.

[3] Dhaundiyal A.B. & Singh S. (2017) Study of distributed activation energy model using various probability distribution functions for the isothermal pyrolysis problem, Rud Zb, 32, 1–14. doi: 10.17794/rgn.2017.4.1.

[4] Miura K. (1995) A new and simple method to estimate f(E) and k0(E) in the distributed activation energy model from three sets of experimental data, Energy & Fuels, 9, 302–307. doi: 10.1021/ef00050a014.

[5] Zhang J., Chen T., Wu J., & Wu J. (2014) Multi-Gaussian-DAEM-reaction model for thermal decompositions of cellulose, hemicellulose and lignin: Comparison of N2 and CO2 atmosphere, Bioresource Technology, 166, 87–95. doi: 10.1016/j.biortech.2014.05.030.

[6] Silva G.F., Camargo F.L., & Ferreira A.L.O. (2011) Application of response surface methodology for optimization of biodiesel production by transesterification of soybean oil with ethanol, Fuel Process Technology, 92, 407–413. doi: 10.1016/j.fuproc.2010.10.002.

[7] Rajendra M., Jena P.C., & Raheman H. (2009) Prediction of optimized pretreatment process parameters for biodiesel production using ANN and GA, Fuel, 88, 868–875. doi: 10.1016/j.fuel.2008.12.008.

[8] Capart R., Khezami L., & Burnham A.K. (2004) Assessment of various kinetic models for the pyrolysis of a microgranular cellulose, Thermochimica Acta, 417, 79–89. doi: 10.1016/j.tca.2004.01.029.

[9] Conesa J.A., Marcilla A., Caballero J.A., & Font R. (2001) Comments on the validity and utility of the different methods for kinetic analysis of thermogravimetric data, Journal of Analytical and Applied Pyrolysis, 58–59, 617–633. doi: 10.1016/S0165-2370(00)00130-3.

[10] Hanaya M., Osawa I., & Pysiak J.J. (2004) Kinetic equations for thermal dissociation processes, Journal of Thermal Analysis and Calorimetry, 76, 521–528. doi: 10.1023/B:JTAN.0000028030.49773.ad.

[11] Yaroshenko A.P. (2005) Theoretical model and experimental study of pore growth during thermal expansion of graphite intercalation compounds, Journal of Thermal Analysis and Calorimetry, 79, 515–519. doi: 10.1007/s10973-005-0571-3.

[12] Criado J.M. & Pérez-Maqueda L.A. (2005) Sample controlled thermal analysis and kinetics, Journal of Thermal Analysis and Calorimetry, 80, 27–33. doi: 10.1007/s10973-005-0609-6.

[13] Burnham A.K. Introduction to chemical kinetics. In: Global chemical kinetics of fossil fuels, Springer International Publishing, Cham, 2017, 25–74.

[14] de Caprariis B., De Filippis P., Herce C., & Verdone N. (2012) Double-Gaussian distributed activation energy model for coal devolatilization, Energy & Fuels, 26, 6153–6159. doi: 10.1021/ef301092r.

[15] Constable F.H. (1925) The mechanism of catalytic decomposition, Proceedings of the Royal Society A: Mathematical, Physical Science, 108, 355–378. doi: 10.1098/rspa.1925.0081.

[16] Vand V. (1943) A theory of the irreversible electrical resistance changes of metallic films evaporated in vacuum, Proceedings of the Physical Society, 55, 222–246. doi: 10.1088/0959-5309/55/3/308.

[17] Pitt G.J. (1962) The kinetics of the evolution of volatile products from coal, Fuel, 267–274.

[18] Hanbaba P. Reaktionkinetische Untersuchungen sur Kohlenwasserstoffenbindung aus Steinkohlen bie niedregen Aufheizgeschwindigkeiten. University of Aachen, 1967.
[19] Anthony D.B. & Howard J.B. (1976) Coal devolatilization and hydrogastification, AIChE Journal American Institute of Chemical Engineers, 22, 625–656. doi: 10.1002/aic.690220403.
[20] Dhaundiyal A. & Singh S. (2017) Asymptotic approximations to the isothermal pyrolysis of deodar leaves using Gamma Distribution, Universitas Scientiarum, 22, 263–284. doi: 10.11144/Javeriana.SC22-3.aatt.
[21] Cai J.M. & Liu R.H. (2007) Parametric study of the nonisothermal n th-order distributed activation energy model involved the Weibull distribution for biomass pyrolysis, Journal of Thermal Analysis and Calorimetry, 89, 971–975. doi: 10.1007/s10973-006-8266-y.
[22] Dhaundiyal A. & Singh S.B. (2017) Approximations to the non-isothermal distributed activation energy model for biomass pyrolysis using the Rayleigh Distribution, Acta Technologica Agriculturae, 20, 78–84. doi: 10.1515/ata-2017-0016.
[23] Dhaundiyal A., Singh S.B., & Hanon M.M. (2018) Application of Archimedean copula in the non-isothermal n^{th} order distributed activation energy model, Biofuels, 1–12. doi: 10.1080/17597269.2018.1442662.
[24] De Filippis P., de Caprariis B., Scarsella M., & Verdone N. (2015) Double distribution activation energy model as suitable tool in explaining biomass and coal pyrolysis behavior, Energies, 8, 1730–1744. doi: 10.3390/en8031730.
[25] Lewellen P.C., Peters W.A., & Howard J.B. (1977) Cellulose pyrolysis kinetics and char formation mechanism, Symposium on Combustion, 16, 1471–1480. doi: 10.1016/S0082-0784(77)80429-3.
[26] Lee C.K., Chaiken R.F., & Singer J.M. (1977) Charring pyrolysis of wood in fires by laser simulation, Symposium on Combustion, 16, 1459–1470. doi: 10.1016/S0082-0784(77)80428-1.
[27] Bamford C.H., Crank J., & Malan D.H. (1946) The combustion of wood, Mathematical Proceedings of the Cambridge Philosophical Society, 42, 166–182. doi: 10.1017/S030500410002288X.
[28] Kansa E.J., Perlee H.E., & Chaiken R.F. (1977) Mathematical model of wood pyrolysis including internal forced convection, Combust Flame, 29, 311–324. doi: 10.1016/0010-2180(77)90121-3.
[29] Dhaundiyal A., Singh S.B., Hanon M.M., & Rawat R. (2018) Determination of kinetic parameters for the thermal decomposition of parthenium hysterophorus, Environmental and Climate Technologies, 22, 5–21. doi: 10.1515/rtuect-2018-0001.
[30] Golikeri S.V. & Luss D. (1972) Analysis of activation energy of grouped parallel reactions, AIChE Journal American Institute of Chemical Engineers, 18, 277–282. doi: 10.1002/aic.690180205.
[31] Burnham A.K. & Dinh L.N. (2007) A comparison of isoconversional and model-fitting approaches to kinetic parameter estimation and application predictions, Journal of Thermal Analysis and Calorimetry, 89, 479–490. doi: 10.1007/s10973-006-8486-1.
[32] Brown M.E., Maciejewski M., Vyazovkin S., et al. (2000) Computational aspects of kinetic analysis: Part A: The ICTAC kinetics project-data, methods and results, Thermochimica Acta, 355, 125–143. doi: 10.1016/S0040-6031(00)00443-3.
[33] Cai J. & Liu R. (2007) Weibull mixture model for modeling nonisothermal kinetics of thermally stimulated solid-state reactions: Application to simulated and real kinetic conversion data, The Journal of Physical Chemistry. B, 111, 10681–10686. doi: 10.1021/jp0737092.
[34] Shaw W.T. & Buckley I.R.C. (2009) The alchemy of probability distributions: beyond Gram-Charlier expansions, and a skew-kurtotic-normal distribution from a rank transmutation map
[35] Aryal G.R. & Tsokos C.P. (2009) On the transmuted extreme value distribution with application, Nonlinear Analysis, Theory, Methods and Applications, 71, e1401–e1407. doi: 10.1016/j.na.2009.01.168.

[36] ELBATAL I., Asha G., & Raja A. (2014) Transmuted exponentiated Fr^echet distribution: Properties and applications, Journal of Statistics Applications and Probability, 3, 379–394. doi: 10.12785/jsap/030309.
[37] Ashour S.K. & Eltehiwy M.A. (2013) Transmuted lomax distribution, American Journal of Applied Mathematics and Statistics, 1, 121–127. doi: 10.12691/ajams-1-6-3.
[38] Aryal G. (2013) Transmuted log-logistic distribution, Journal of Statistics Applications and Probability, 2, 11–20. doi: 10.12785/jsap/020102.
[39] Elbatal I. (2013) Transmuted modified inverse weibull distribution: A generalization of the modified inverse Weibull probability distribution, International Journal of Mathematical Archive, 4, 117–129.
[40] Elbatal I. & Aryal G. (2013) On the transmuted additive Weibull distribution, Austrian Journal of Statistics, 42, 117–132. doi: 10.17713/ajs.v42i2.160.
[41] Abd El-Monsef M.M.E. (2016) A new Lindley distribution with location parameter, Communications in Statistics -- Theory Methods, 45, 5204–5219. Doi: 10.1080/03610926.2014.941496.
[42] Dhaundiyal A. & Singh S.B. (2016) Asymptotic approximations to the distributed activation energy model for non isothermal pyrolysis of loose biomass using the Weibull distribution, Archivum Combustionis, 36, 131–146.
[43] Chatterjee A. & Chatterjee A. (2012) Use of the Fréchet distribution for UPV measurements in concrete, NDT & E International : Independent Nondestructive Testing and Evaluation, 52, 122–128. doi: 10.1016/j.ndteint.2012.07.003.
[44] Dhaundiyal A. & Singh S.B. (2017) Parametric Study of NTH Order distributed activation energy model for isothermal pyrolysis of forest waste using Gaussian distribution, Acta Technologica Agriculturae, 20, 23–28. doi: 10.1515/ata-2017-0005.
[45] Dey S., Dey T., & Kundu D. (2014) Two-parameter Rayleigh Distribution: Different methods of estimation, American Journal of Mathematical and Management Sciences, 33, 55–74. doi: 10.1080/01966324.2013.878676.
[46] Ungerer P. & Pelet R. (1987) Extrapolation of the kinetics of oil and gas formation from laboratory experiments to sedimentary basins, Nature, 327, 52–54. doi: 10.1038/327052a0.
[47] Burnham A.K. & Braun R.L. (1999) Global kinetic analysis of complex Materials, Energy & Fuels, 13, 1–22. doi: 10.1021/ef9800765.
[48] Burnham A.K., Braun R.L., Gregg H.R., & Samoun A.M. (1987) Comparison of methods for measuring kerogen pyrolysis rates and fitting kinetic parameters, Energy & Fuels, 1, 452–458. doi: 10.1021/ef00006a001.
[49] Bourguignon M., Ghosh I., & Cordeiro G.M. (2016) General Results for the transmuted family of distributions and new models, Journal of Probability and Statistics, 2016, 1–12. doi: 10.1155/2016/7208425.
[50] Dhaundiyal A., Singh S.B., Atsu D., & Dhaundiyal R. (2019) Application of Monte Carlo simulation for energy modeling, Journal of American Chemical Society Omega, 4(3), 4984–4990.
[51] Dhaundiyal, A. and Toth, L. (2020) 'Modeling of Hardwood Pyrolysis Using the Convex Combination of the Mass Conversion Points', Journal of Energy Resources Technology, Transactions of the ASME. doi: 10.1115/1.4045458.
[52] Dhaundiyal, A., Singh, S.B. The generalisation of a multi-reaction model for polynomial ramping of temperature. J Therm Anal Calorim (2020). https://doi.org/10.1007/s10973-020-09650-7.

Appendices

To demonstrate the scheme, the Laplace method of integration is chosen to simplify the DAEM problem of iterative computation of double integrals. This method is useful to solve the integral of the form

$$I = \int_a^b e^{-BQ(x)} s(x)\, dx$$

where a and b can be finite or infinite. Here $Q(x)$ and $s(x)$ are replaced by their local expansion when $x = x_0$. The same concept is used for approximating the integral form as mentioned earlier.

The general solution of Laplace method for interior points is given by

$$I = 1.7724 e^{-BQ(x_0)} s(x_0) \sqrt{\frac{1}{|BQ''(x_0)|}}$$

Note: $s(x_0) \neq 0$

However, $c(y_e) \ll 1$, so its contribution to integral is negligibly small for interval of $0 < y_e < 1$.

A. The Lindley (three-parameter) distribution

Table A.1: Kinetic parameters for the Lindley distribution functions.

Average activation energy (kJ/mol)	Frequency factor (min^{-1})		R^2	($p(E) \to 0$) After transmutation		($E \to \infty$) After transmutation	
				τ	R^2	τ	R^2
1.0543×10^4	A_1	2×10^{14}	0.95	0.0008	0.93	0.05	0.94
				0.02	0.94	0.04	0.95
1.4906×10^4	A_2	2×10^{20}	0.93	0.0008	0.94	0.05	0.94
				0.02	0.94	0.04	0.93
2.7733×10^3	A_3	2×10^3	0.96	0.0008	0.97	0.05	0.97
				0.02	0.97	0.04	0.98

$$p(E) = \frac{\theta^2}{\theta + \psi}(1 + \psi(E - \gamma))e^{-\theta(E-\gamma)}, E > \gamma \geq 0, \theta + > 0 \tag{A.1}$$

This can be solved through the convex combination of two different functions, but provided that $\forall f_1(E), f_2(E) \varepsilon C$ and $C \subseteq \mathbb{R}$

Equation (A.1) can be rewritten as

$$p(E) = kf_1(E) + (1-k)f_2(E) \tag{A.2}$$

where k is the mixing fraction and φ is a scale parameter

$$k = \frac{\theta}{\theta+\psi} \quad \varphi = \frac{1}{\theta} \text{ and } \omega_1 = \frac{\gamma}{\varphi}$$

The mean and the variance of the Lindley distribution are E_{01} and $\sigma^2{}_1$, respectively

$$E_{01} = \frac{\theta(1+\gamma\theta)+\psi(2+\gamma\theta)}{\theta(\theta+\psi)} \quad \sigma^2{}_1 = \frac{[\{\theta(2+\gamma\theta(2+\gamma\theta))+\psi(6+\gamma\theta(4+\beta\theta))\}]}{\theta^2(\theta+6)}$$

Energy is rescaled as

$$y_1 = \frac{E}{\gamma}, \quad y_{s1} = \frac{E_s}{\gamma}, \text{ and } y_{w1} = \frac{E_w}{\gamma}$$

The expressions for $f_1(E)$ and $f_2(E)$ functions are

$$f_1(E) = \theta\psi(E-\gamma)\exp\{-\theta(E-\gamma)\} \tag{A.3}$$

$$f_2(E) = \theta^2\exp\{-\theta(E-\gamma)\} \tag{A.4}$$

The simplified form obtained after incorporating the energy rescaling

$$f_1(y_1) = \omega_1\psi(y_1-1)\exp\{-\omega_1(y_1-1)\}$$

$$f_2(y_1) = \theta^2\exp\{-\omega_1(y_1-1)\}$$

The resultant form of density function in terms of energy rescaling factor is

$$p(y_1) = kf_1(y_1) + (1-k)f_2(y_1)$$

Note: The whole integral transforms into convex combination of Gamma functions

$$\left(\Gamma(\psi, \omega_1 y_1), \Gamma\left(\theta^2, \omega_1 y_1\right)\right) \text{ if } \left(\frac{y_s - y_1}{y_w}\right) \to \infty$$

B. Weibull distribution (three parameters)

Table B.1: Kinetic parameters for the Weibull distribution functions.

Average activation energy (kJ/mol)	Frequency factor (min^{-1})		R^2	($p(E) \to 0$) After transmutation		($E \to \infty$) After transmutation	
				τ	R^2	τ	R^2
89	A_1	10^8	0.920	0.0008	0.94	0.05	0.90
				0.02	0.96	0.04	0.91
99	A_2	10^9	0.910	0.0008	0.94	0.05	0.89
				0.02	0.95	0.04	0.90
69	A_3	10^7	0.900	0.0008	0.93	0.05	0.88
				0.02	0.95	0.04	0.89

$$p(E) = \frac{\beta}{\eta}\left(\frac{E-\gamma}{\eta}\right)^{(\beta-1)} e^{-\left(\frac{E-\gamma}{\eta}\right)^{\beta}} \tag{B.1}$$

This is the expression for the Weibull distribution function for three parameters.

Energy rescaling factor is given as follows:

$$y_2 = \frac{E}{\gamma},\ y_{s2} = \frac{E_s}{\gamma} \text{ and } y_{w2} = \frac{E_w}{\gamma} \tag{B.2}$$

After simplifying (B.1.), the required form of density function $p(y_2)$ is

$$p(y_2) = \frac{\beta}{\eta}\omega_2{}^{\beta-1}(y_2-1)^{\beta-1}e^{-\left((\omega_2(y_2-1))^{\beta}\right)}$$

where

$$\omega_2 = \frac{\gamma}{\eta}$$

The mean and the variance of the Weibull distribution are E_{02} and $\sigma^2{}_2$, respectively:

$$E_{02} = \gamma + \eta\Gamma\left(\frac{1}{\beta}+1\right) \text{and}\, \sigma^2{}_2 = \eta^2\left[\Gamma\left(\frac{2}{\beta}+1\right) - \left\{\Gamma\left(\frac{1}{\beta}+1\right)\right\}^2\right]$$

C. Gaussian distribution

Table C.1: Kinetic parameters for the Gaussian distribution functions.

Average activation energy (kJ/mol) E	Frequency factor (min^{-1}) (A)		R^2	$(p(E) \to 0)$ After transmutation		$(E \to \infty)$ After transmutation	
				τ	R^2	τ	R^2
120	A_1	2×10^{10}	0.9315	0.0008	0.93	0.05	0.97
				0.02	0.92	0.04	0.94
119.6	A_2	2×10^{7}	0.9308	0.0008	0.93	0.05	0.94
				0.02	0.92	0.04	0.94
150	A_3	2×10^{13}	0.9307	0.0008	0.93	0.05	0.96
				0.02	0.92	0.04	0.95

The required expression for Gaussian distribution is given as follows:

$$p(E) = \frac{1}{\sqrt{2\pi\sigma^2}} e^{-\left(\frac{E-E_0}{\sqrt{2}\sigma}\right)^2} \tag{C.1}$$

The energy correction factor for the Gaussian distribution is expressed as

$$y_3 = \frac{E}{E_0}, \quad y_{s3} = \frac{E_s}{\gamma} \text{ and } y_{w3} = \frac{E_w}{\gamma} \tag{C.2}$$

Simplifying eq. (C.1) through incorporating energy factors, we have

$$p(y_3) = \frac{1}{\sqrt{2\pi\sigma^2}} \exp\left(-\left(\frac{E_0}{\sqrt{2}\sigma}\right)^2 (y_3 - 1)^2\right) \tag{C.3}$$

Putting $\omega_3 = \left(\frac{E_0}{\sqrt{2}\sigma}\right)^2$ in eq. (C.3), the required form is

$$p(y_3) = \frac{1}{\sqrt{2\pi\sigma^2}} \exp\left(-\omega_3 (y_3 - 1)^2\right)$$

D. Frechet distribution

Table D.1: Kinetic parameters for the Frechet distribution function.

Average activation energy (kJ/mol)	Frequency factor (min^{-1})		R^2	($p(E) \to 0$) After transmutation		($p(E) \to 0$) After transmutation	
				τ	R^2	τ	R^2
119.6	A_1	2×10^{10}	0.94	0.008	0.93	0.05	0.96
				0.02	0.92	0.04	0.95
140	A_2	2×10^{12}	0.80	0.008	0.80	0.05	0.79
				0.02	0.81	0.04	0.80
171	A_3	2×10^{15}	0.15	0.008	0.16	0.05	0.13
				0.02	0.97	0.04	0.14

$$p(E) = \frac{\rho}{\delta}\left(\frac{E-\mu}{\delta}\right)^{-(1+\rho)} e^{-\left(\frac{E-\mu}{\delta}\right)^{-\rho}} \tag{D.1}$$

Equation (D.1) represents the density function of the Frechet distribution.

The energy factor for distribution is given by

$$y_4 = \frac{E}{\mu},\ y_{4S} = \frac{E_S}{\mu} \text{ and } y_{4S} = \frac{E_W}{\mu} \tag{D.2}$$

Introducing the energy rescaling factors in eq. (D.1), we have

$$p(y_4) = \frac{\rho}{\delta}\left(\frac{\mu(y_4-1)}{\delta}\right)^{-(1+\rho)} e^{-\left(\frac{\mu(y_4-1)}{\delta}\right)^{-\rho}} \tag{D.3}$$

Putting $\omega_4 = \frac{\mu}{\delta}$ in eq. (D.3), the desired expression will be

$$p(y_4) = \frac{\rho}{\delta}\omega_4{}^{-(\rho+1)}(y_4-1)^{-(1+\rho)}e^{-\{\omega_4(y_4-1)\}^{-\rho}}$$

The mean and the variance of the Frechet distribution are E_{04} and $\sigma^2{}_4$, respectively,

$$E_{04} = \mu + \eta\Gamma\left(1-\frac{1}{\rho}\right)(\rho>1)$$

$$\sigma^2{}_4 = \eta^2\left[\Gamma\left(1-\frac{2}{\beta}\right) - \left\{\Gamma\left(1-\frac{1}{\beta}\right)\right\}^2\right](\rho>2)$$

E. Rayleigh distribution

Table E.1: Kinetic parameters for the Rayleigh distribution functions.

Average activation energy (kJ/mol)	Frequency factor (min^{-1})		R^2	($p(E) \to 0$) After transmutation		($E \to \infty$) After transmutation	
				τ	R^2	τ	R^2
127	A_1	2×10^8	0.94	0.0008	0.940	0.05	0.95
				0.02	0.935	0.04	0.96
79	A_2	2×10^6	0.93	0.0008	0.941	0.05	0.94
				0.02	0.937	0.04	0.95
234	A_3	2×10^{21}	0.93	0.0008	0.930	0.05	0.95
				0.02	0.92	0.04	0.94

The density function of two parameters (2-P) is expressed as

$$p(E) = 2\chi(E - \phi)\exp\left\{-\chi(E - \phi)^2\right\} \tag{E.1}$$

The energy factor for the Rayleigh distribution is expressed as

$$y_5 = \frac{E}{\phi}\ \ y_{s5} = \frac{E_s}{\phi}\ \text{and}\ y_{w5} = \frac{E_w}{\phi}$$

Rearranging eq. (E.1), we will have the required form of density function:

$$p(y_5) = 2\frac{\omega_5}{\phi}(y_5 - 1)\exp\left\{-\omega_5(y_5 - 1)^2\right\} \tag{E.2}$$

where

$$\omega_5 = \frac{\phi^2}{\sigma^2{}_5\left(1 - \frac{\pi}{4}\right)}$$

The mean and the variance of the Rayleigh distribution are E_{05} and $\sigma^2{}_5$, respectively,

$$E_{05} = \phi + \frac{\sqrt{\pi}}{2\chi}\ \text{and}\ \sigma^2{}_5 = \frac{1}{\chi} - \frac{\pi}{4\chi}$$

where ϕ kJ/mol is the location parameter

The physical significance of location parameter is the minimum energy required to initiate the chemical reaction.

Shruti Tikhe, Kanchan Khare, S. N. Londhe

6 Air quality forecasting using soft computing techniques

Abstract: Air pollution is a fundamental problem among the critical challenges of modern society. Acute as well as long-term exposure may pose serious health problems. Real-time air quality predictions are necessary to forecast the pollutant concentrations. This information can be used to issue early air quality warnings that allow government and people to take preventive measures.

The authors offer application of soft computing techniques of artificial neural networks (ANN) and genetic programming (GP) for air quality parameter forecasting, a few time steps in advance. Seasonal forecasting models are also useful for proactive measures. Availability of continuous field data sometimes is a challenge. It is time-consuming and expensive. Modeling options are suggested for situations when there are constraints about data availability and the urgent need to understand and broadcast the air quality.

A case study of Pune, a metropolitan city in India, demonstrating the strengths of GP is presented. We have evaluated the models by a correlation coefficient, *d*-statistics, root mean squared error, and mean bias error.

Keywords: Air quality, soft computing, ANN, GP

6.1 Introduction

Air is a fundamental need for survival and development of all lives on the Earth. It has major impact on health and also influences the economic development of the country [18]. Air quality describes the chemical state of the atmosphere at a particular time and place. Rapid industrialization, development of cities, and concerned human activities are the major causes for an alarming increase in air pollution levels and degradation of the air quality especially in almost all the metro cities of the world [23]. Short-term exposure to high levels of pollution may pose severe health impacts such as eye irritation, difficulty in breathing, and cardiovascular and pulmonary health effects. Health concerns associated with long-term exposure can be

Shruti Tikhe, DTK Hydronet Solutions, Bavdhan, Pune, Maharashtra State, India
Kanchan Khare, Civil Engineering Department, Symbiosis Institute of Technology Lavale, Pune, Maharashtra State, India
S. N. Londhe, Department of Civil Engineering, Vishwakarma Institute of Information Technology Kondhwa, Pune, Maharashtra State, India

https://doi.org/10.1515/9783110628616-006

listed as the threat of cancer, risk of premature death, and also damage to the body's immune, respiratory, neurological, and reproductive systems. Senior citizens, kids, and individuals with preexisting heart and lung maladies along with diabetics are at greater risk for air pollution-related health effects [19]. The World Health Organization has issued guidelines related to permissible levels of pollutants with an objective to protect human health and the environment. In spite of significant progress in understanding the process of emission and fates of the air pollutants as well as in decreasing their ambient levels in urban localities for the last five decades, air pollution is estimated to destroy the lives of about 4 million people every year worldwide [34]. It is therefore worthwhile to model and forecast the air quality.

6.2 Issues and challenges

Air quality is described by the concentrations of major air pollutants such as sulfur dioxide, nitrogen oxides, and respirable suspended particulate matter (SO_2, NO_x, and RSPM) that reflect the status of air pollution in the environment [36]. Air quality assessment studies are essential to determine the trends and effects of air pollutants on human lives, animals, soil, and environment. Source contribution to pollution can also be identified and accordingly controlling strategies can be decided [11].

Air pollution comprises a mixture of gases and particles in harmful amounts that are released into the atmosphere due to either natural or human activities. The leading sources of air pollution are natural as well as anthropogenic. Natural phenomena such as volcanic eruptions and forest fires release harmful substances such as SO_2, CO_2, NO_2, and CO.

Anthropogenic sources include the pollutants entering from a fixed point, mobile sources, and area sources in addition to the burning of fossil fuel. Pollutants emitted from human-made sources include hydrogen, oxygen, nitrogen, sulfur, metal compounds, and particulate matter.

They affect badly on human health and the environment. The global impact of air pollution is on climate change and particularly on global warming.

During the last five decades, scientific research has been carried out to model as well as forecast the air quality. Various methods and techniques have been evolved for modeling of air pollutants. For effective air quality management, monitoring and record-keeping of the pollutant concentration data become a vital prerequisite. Many countries have their real-time air quality monitoring as well as forecasting systems. Developed countries have their established air quality monitoring system, but the issue is with the developing countries where availability of the continuous data is a significant challenge.

6.3 Air quality modeling

It has been observed that air quality assessment and modeling has been carried out using a variety of approaches. The complexity of the problem and the availability of the resources decide the choice of method [6]. Traditionally, hard computing approaches have been used to carry out air quality modeling and forecasting. They include either physical, deterministic, stochastic, statistical, or numerical models [30]. Deterministic models require emission inventory and meteorological variables to calculate the pollutant concentrations. These models have been used traditionally to calculate air pollutants in urban areas [20].

Physical models are costly, data-intensive, and challenging to construct and operate; as a result, their use has been restricted only to understand the complex urban dispersion phenomenon [1]. Majority of the dispersion models developed to evaluate and predict the pollutants in urban areas are "causal" in nature and therefore fail to predict the "extreme" concentrations [9]. In air quality management, forecasting and broadcasting of the peak concentrations (extremes) are very important to give sufficient warning to the citizens. Statistical techniques represent only a specific monitoring station; as a result, they cannot be extended to other regions with different meteorological conditions [14]. The interrelationship between human, climate, and air pollution is extremely complex; as a result, it is very difficult to be represented in deterministic models without developing a separate statistical model [6]. As meteorology affects air quality and this relationship is complex and nonlinear, data-driven approaches are suitable; however, they are also site-specific. They provide tools to facilitate the conversion of the data sets into a convenient form and convey a better understanding of the processes hidden in these data sets by expressing the mathematical statement of the relations between various parameters in the data [2]. Soft computing tools have been developed as an alternative to deterministic and physical models.

6.4 Forecasting of air quality parameters

Air quality parameters are stochastic in nature; thus, short-term predictions are possible based on past meteorological data [22]. Soft computing tools are modern forecasting approaches. Resource availability and accuracy requirements decide the choice of the tool. Chosen tool should be robust enough to take into account the data fluctuations without compromising the accuracy. Extensive experimentation is required to identify such a tool [12].

6.4.1 Artificial neural networks

There are many data-driven approaches known in computer science today. One of the most popular approach is artificial neural networks (ANNs). ANNs are biologically inspired, parallel computational tools. They consist of simple highly interconnected processing elements that process the input similar to the human brain [28]. ANN is broadly classified as a feed-forward neural network and a recurrent neural network. Figure 6.1 shows a feed-forward network [15], which typically has three layers, namely input, output and hidden layer. Input layer neurons receive the input variables from the problem, and output layer neuron/neurons yield the network output (desired result). ANN works similar to the human brain on two fronts. First one is acquiring knowledge from the input data via the network of neurons through a learning process and interneuron connection strengths are utilized to store the acquired knowledge. The weights [between the input layer and hidden layer as well as hidden layer and output layer] represent the relative strength of specific input parameters in the computation of output. An artificial neuron weighs and sums up the inputs and compares the results with a predefined threshold. Iterations are carried out until the stopping criteria are reached. The network-generated output heavily depends on the type of activation functions (also called transfer or squashing function) used. Three types of transfer functions are generally used, namely, step, ramp, and sigmoid function. The sigmoid function is the most

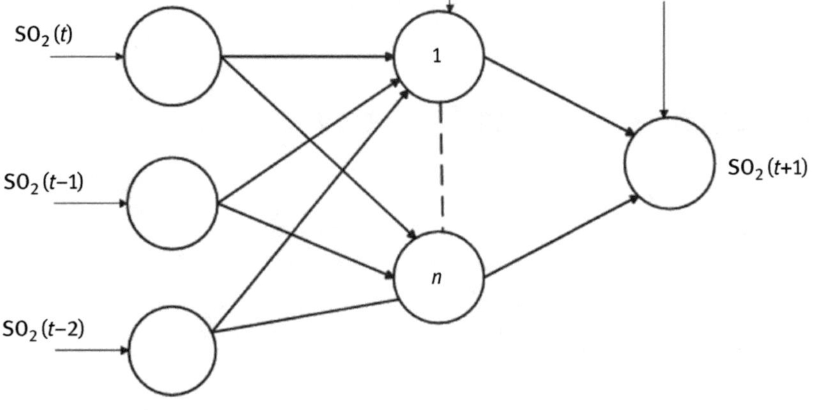

Figure 6.1: The feedforward network.
Source: The ASCE Task Committee, 2000 [15].

commonly used transfer function for air quality modeling [13]. The recurrent network has a feedback path which makes it sequential rather than combinatorial. Once cyclic connections are included, the network becomes a nonlinear dynamic system. Backpropagation is one of the most widely used algorithms in the feedforward type of networks, which is based on the steepest descent method.

The performance of ANN models entirely depends on the quality and the quantity of the data. ANNs were considered as black box earlier but some researchers have shown that the network incorporates all of the physics underlying the process since they are embedded in the data and ANN is not merely playing with numbers [25]. Similarly, the works of Londhe and Shah [10] and Londhe and Panchang [26] throw a light on the work of knowledge extraction from the trained network. Deciding ANN architecture is a complex process that demands the skill of the modeler. There is no fixed rule to determine the data division and the number of neurons in the hidden layer. It has been reported that ANN models may suffer from overfitting due to the addition of too many hidden neurons or executing a large number of iterations. Some researchers have found that ANNs have a capability of prediction of extreme events when combined with other techniques such as wavelets [4].

6.4.2 Genetic programming

Genetic programming (GP) is an evolutionary data-driven modeling approach that is similar to genetic algorithms (GA) and follows the Darwinian principle of the survival of the fittest and obtains the solution to the problem through the process of crossover, mutation, and reproduction. GP can be conveniently used as a regression tool as the solution of GP is a computer program or an equation as against a set of numbers in GA [24].

The process of GP starts with a population of randomly generated computer program initially constructed from the data set on which genetic operations are performed using function and terminal set. Function set comprises the operator to be used such as addition, subtraction, logarithm, and square root. Terminal set consists of values such as inputs, constants, and temporary variables on which the function set operates. From the population so generated, four programs are randomly selected, and a tournament is conducted.

GP measures the performance of each program, and the two programs are selected based on performance. GP algorithm copies the two winner programs and transforms these copies into two new programs via crossover and mutation as per fitness [17]. Creation of offsprings continues iteratively till a specified number of offsprings are created in the generation, and until a specified number of generations are produced. The resulting offspring at the end of the process is either a computer program or an equation. It is the solution to the problem. The generated equation or program can be directly applied to unseen data to obtain required predictions [25].

Figure 6.2 shows a typical GP flowchart. GP can evolve a formula or an equation by automatically selecting the significant inputs contributing to the model development and simultaneously ignoring the variables with the least contribution to the model. Three forms of GP are available, namely standard or tree-based GP, linear GP (LGP), and automatic induction of machine (AIM) code GP. Tree-type GP operates on parse trees, which is constituted by a function set and a terminal set. Crossover operation is carried out by randomly swapping the subtrees between the selected individuals [2]. The typical GP process is continued until the stopping criterion is met. LGP evolves sequences of instructions using an "imperative programming language" or a machine [8]. In LGP, the expressions of a functional programming language (like LISP) are substituted by programs of imperative language (like C/C++) [3]. Noneffective codes (introns) are identified and eliminated. It allows the corresponding useful instructions to be extracted from a program, and the process can be accelerated [7]. AIMGP is a particular form of LGP. For the given dataset, AIMGP writes a computer program and results with the "Best Program" as well as "Best Team." It creates a total of 30 programs in a project and produces five teams of 1,3,5,7,9 program combinations [5].

6.5 Case study in India

In the present case study, one day and two days ahead air pollutant forecasting models have been developed for Pune city (18.5218.5204°N, 73.8567°E), which is one of the highly polluted metropolitan cities located in the Maharashtra state of India. It is the premier industrial center and the fastest developing city of the country [37]. The city is located on the western side of the Maharashtra state on the Deccan plateau at an elevation of 560 m above mean sea level at Karachi. Rapid industrialization along with boom in construction sector has resulted in exponential vehicular population growth, which has resulted in increased pollutant concentrations and worsening of the air quality of Pune for the last few years [32]. Central Pollution Control Board (CPCB) of India has declared SO_2, NO_x, and RSPM as the criteria air pollutants which decide the air quality [35]. The national ambient air quality standards of India sets the maximum permissible limit for daily average concentrations of SO_2, NO_x, and RSPM as 80, 80, 100 µg/cum, respectively, for residential urban areas. For the last decade, it has been observed that the upper limits have been crossed for SO_2, NO_x, and RSPM with a maximum value recorded as high as 87,158 and 680 µg/cum, respectively, for the study area [33].

Daily values of the pollutant concentrations recorded by Maharashtra pollution control board from January 2005 to July 2015 were used for the development of annual and seasonal models. In order to obtain the complete annual picture about the air quality of Pune, the annual models are developed by arranging the yearly data

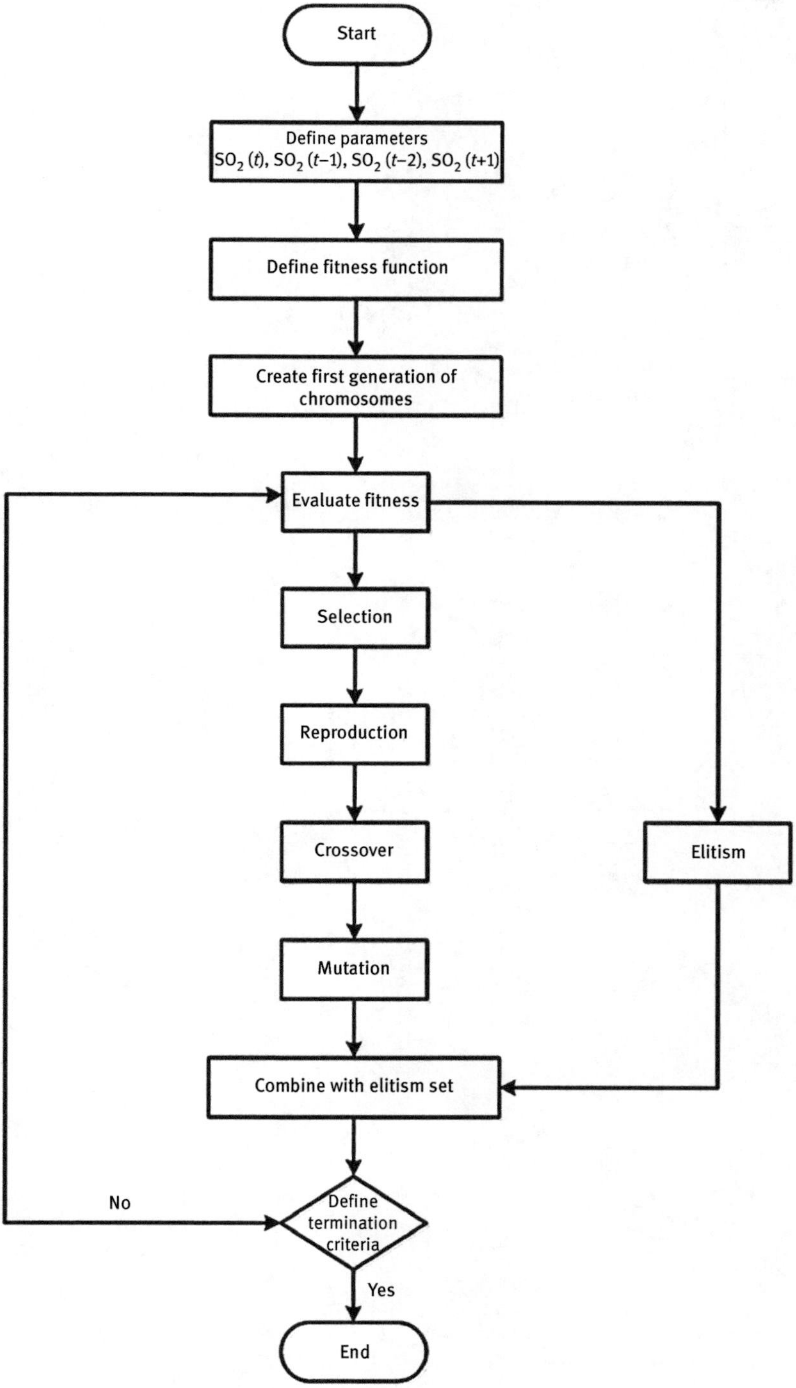

Figure 6.2: Typical GP Flowchart.
Source: Yadav & Sharma, 2019.

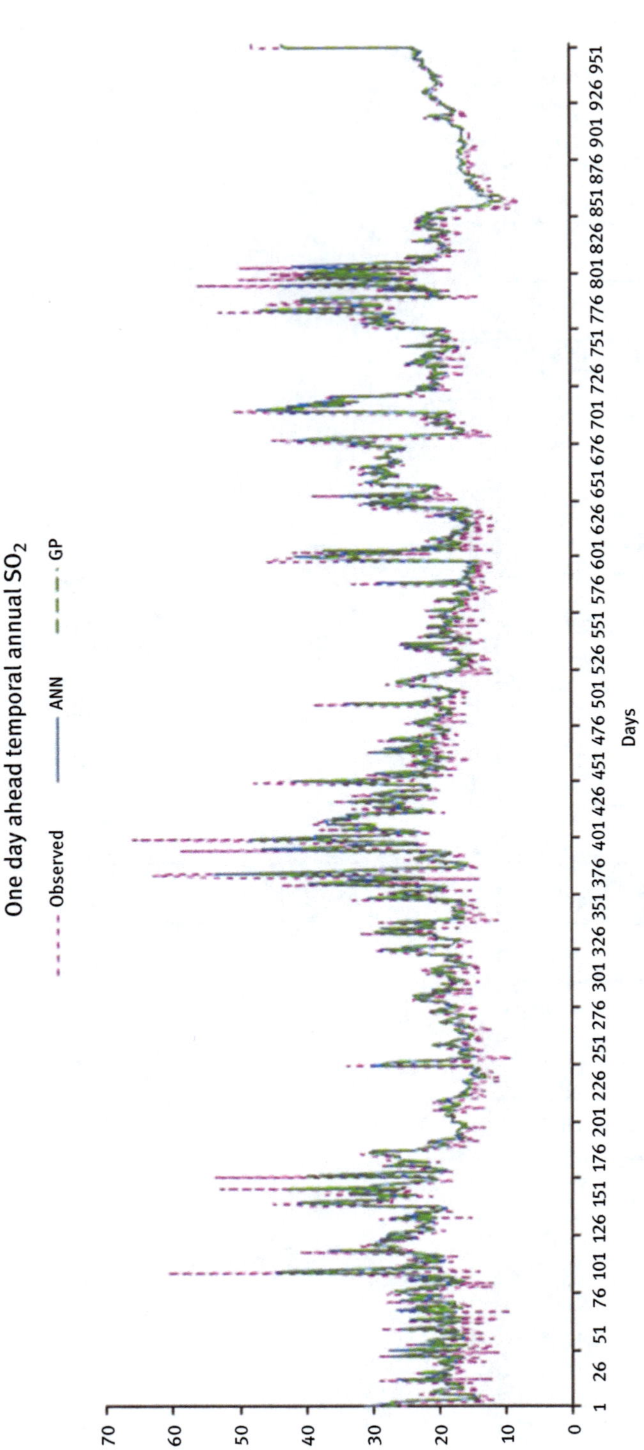

Figure 6.3: Time series plot for 1 day ahead temporal annual SO_2.

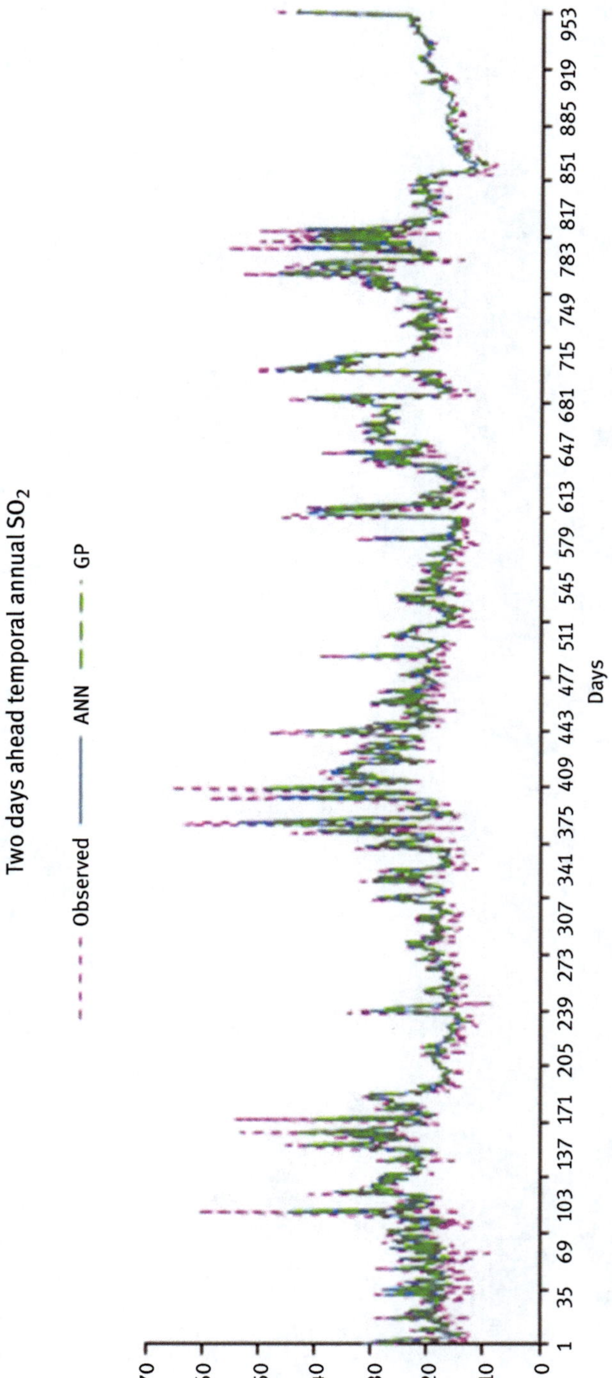

Figure 6.4: Time series plot for 2 days ahead temporal annual SO_2.

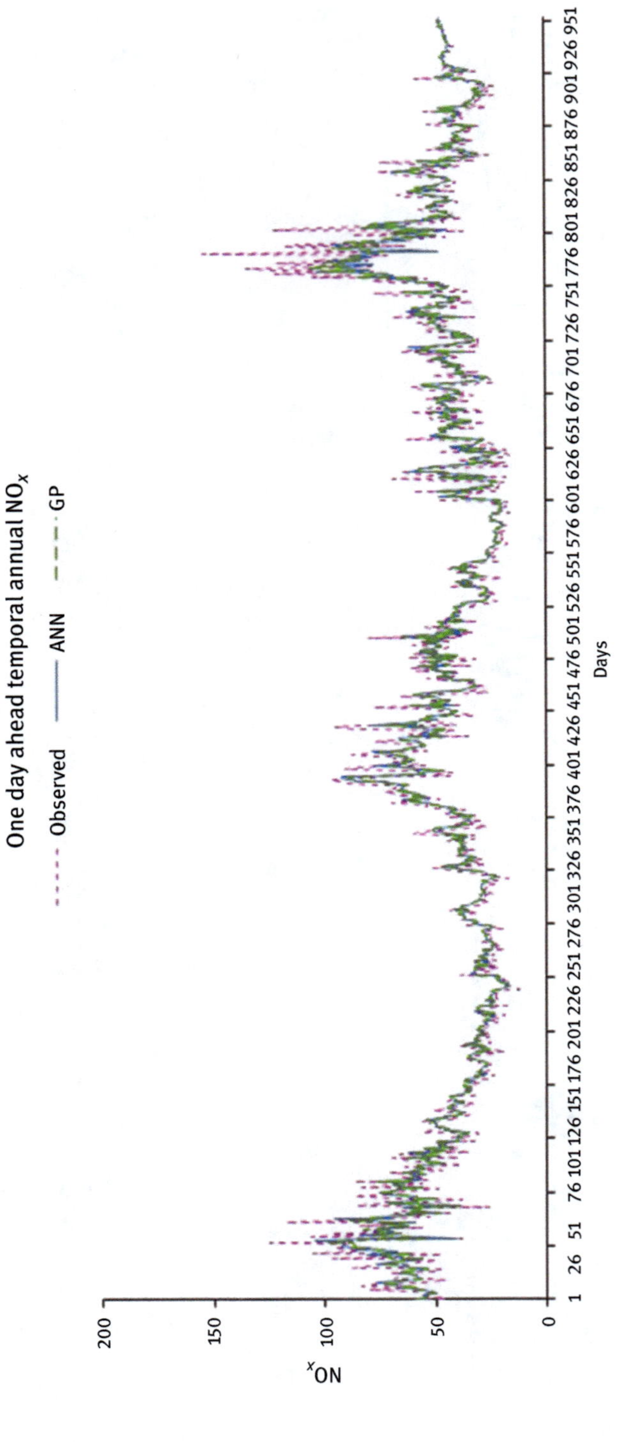

Figure 6.5: Time series plot for 1 day ahead temporal annual NO_x.

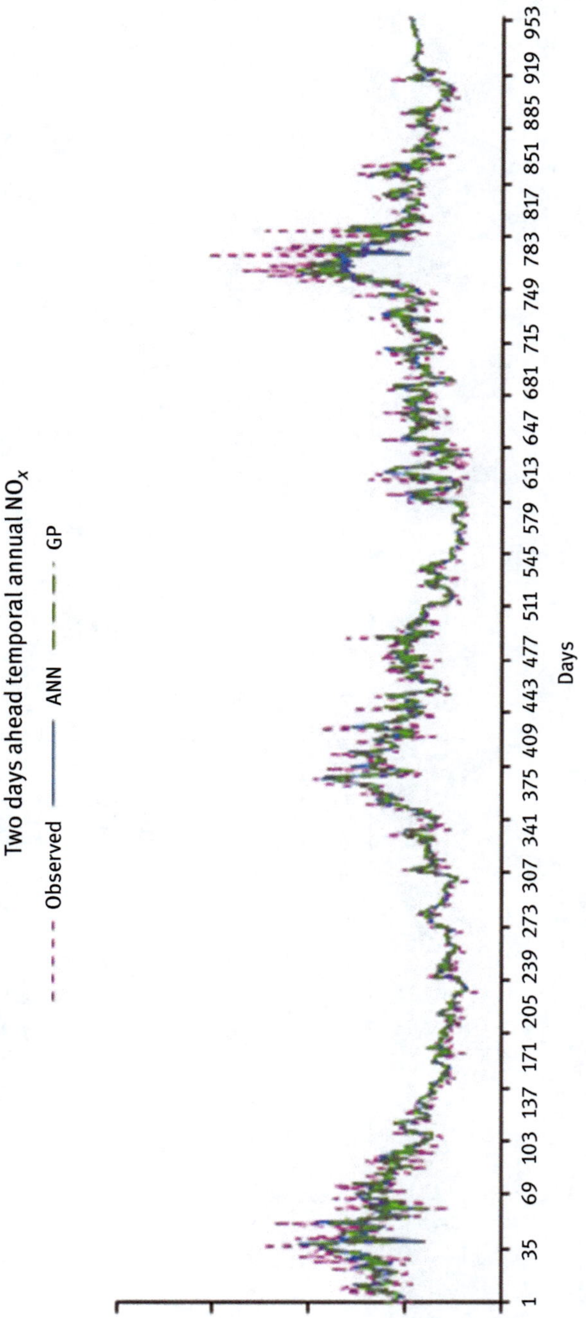

Figure 6.6: Time series plot for 2 days ahead temporal annual NO_x.

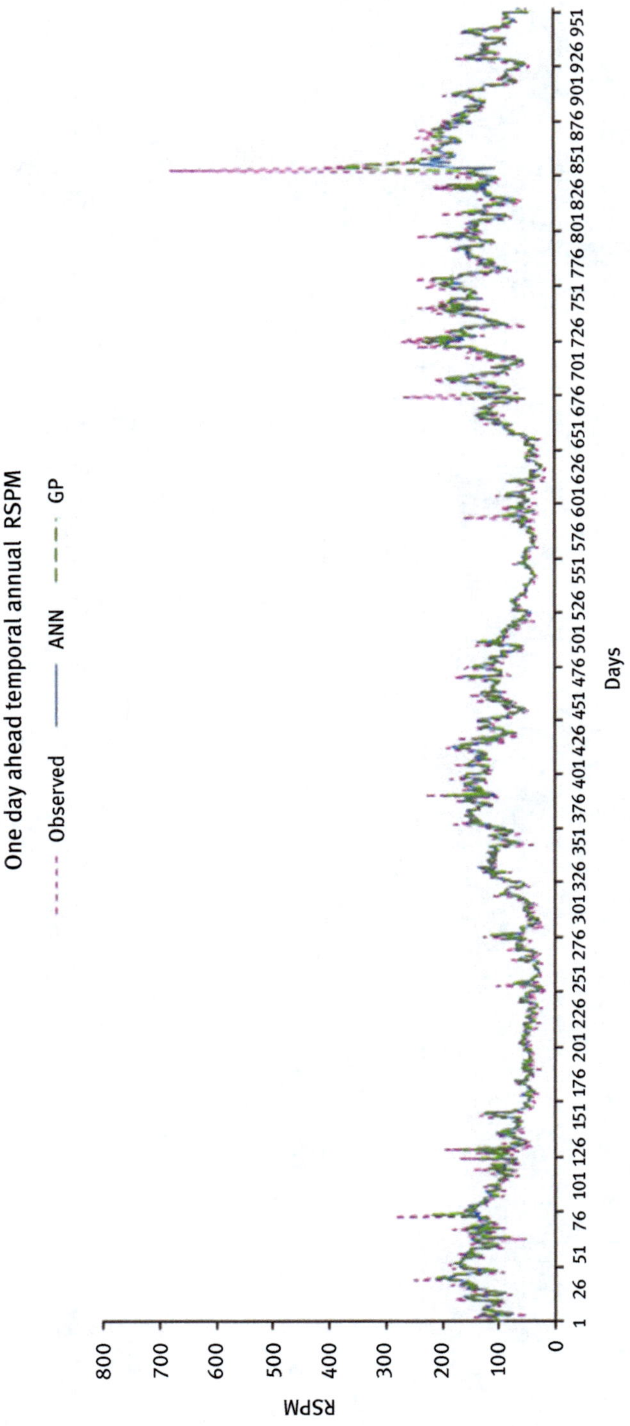

Figure 6.7: Time series plot for 1 day ahead temporal annual RSPM.

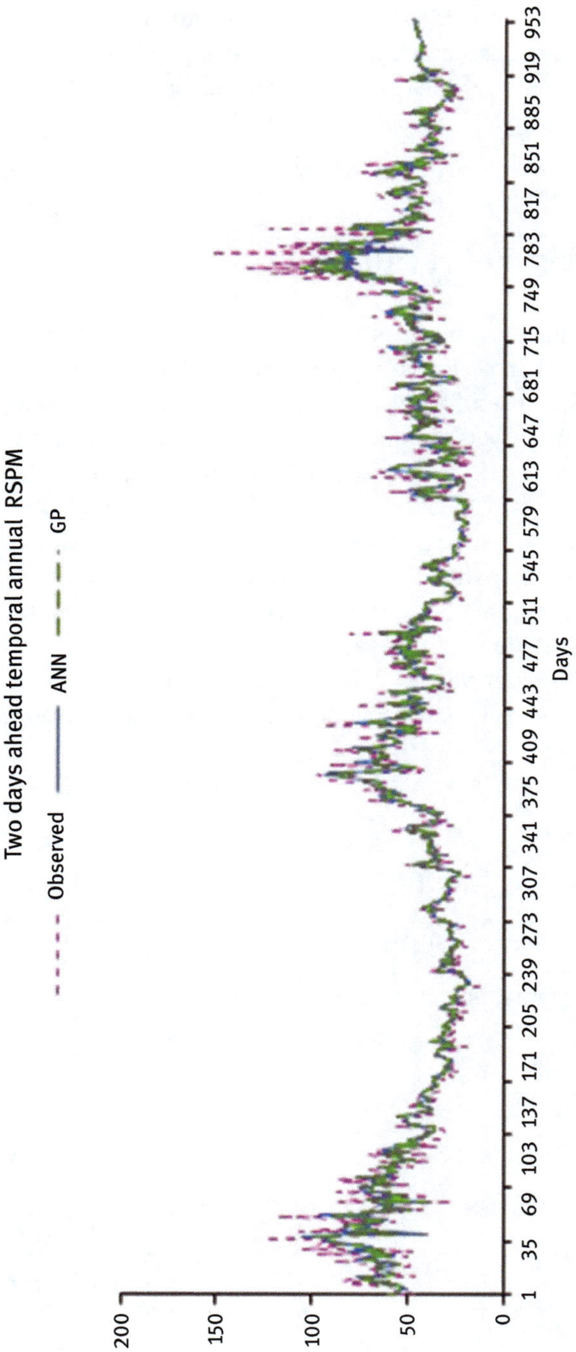

Figure 6.8: Time series plot for 2 days ahead temporal annual RSPM.

serially [16]. Similarly, pollutant concentrations vary with the environmental conditions as observed in different seasons. In India, four seasons such as winter (Jan–Feb), summer (Mar–May), monsoon (Jun–Sept), and post monsoon (Oct–Dec) are observed [24]. Seasonal models are also developed by dividing the data into different seasons.

6.5.1 Experimentation using ANN and GP

Air quality is time-dependent phenomenon. In this chapter, using soft computing approaches of ANN and GP, annual and seasonal models have been developed with the time series data of the criteria air pollutant concentrations. Uniform seasonal patterns and linearity between the inputs were assumed while developing all the forecasting models [26]. The data pertaining to the pollutant concentrations (SO_2, NO_x, and RSPM) recorded from January 2005 to July 2015 have been considered for model development. The input optimization has been carried out using correlation analysis, considering the annual temporal data spanning over the week. Inputs were added one by one, and the multiple linear regression (MLR) models were developed and evaluated. The correlation coefficient of the MLR models suggested that three, four, and three predecessor values shall be used as model inputs for SO_2, NO_x, and RSPM models, respectively. The same number of inputs was used for respective seasonal models as well.

Temporal annual 1 day ahead forecasting models can be written as [29]

$$SO_2(t+1) = f(SO_2(t), SO_2(t-1), SO_2(t-2)) \tag{6.1}$$

$$NO_x(t+1) = f(NO_x(t), NO_x(t-1), NO_x(t-2), NO_x(t-3)) \tag{6.2}$$

$$\text{RSPM}\,(t+1) = f(\text{RSPM}\,(t), \text{RSPM}\,(t-1), \text{RSPM}\,(t-2)) \tag{6.3}$$

Temporal annual 2 days ahead forecasting models can be written as

$$SO_2(t+2) = f(SO_2(t), SO_2(t-1), SO_2(t-2)) \tag{6.4}$$

$$NO_x\ (t+2) = f(NO_x\ (t), NO_x\ (t-1), NO_x\ (t-2), NO_x\ (t-3)) \tag{6.5}$$

$$\text{RSPM}\,(t+2) = f(\text{RSPM}\,(t), \text{RSPM}\,(t-1), \text{RSPM}\,(t-2)) \tag{6.6}$$

and temporal seasonal 1 day ahead forecasting models can be written as

$$\begin{aligned} &SO_2(t+1) = f(SO_2(t), SO_2(t-1), SO_2(t-2)) \\ &\qquad \text{for winter/summer/}monsoon/post\text{ monsoon} \end{aligned} \tag{6.7}$$

$$\begin{aligned} &NO_x\ (t+1) = f(NO_x\ (t), NO_x\ (t-1), NO_x\ (t-2), NO_x\ (t-3)) \\ &\qquad \text{for winter/summer/ monsoon/postmonsoon} \end{aligned} \tag{6.8}$$

$$\text{RSPM}\,(t+1) = f(\text{RSPM}\,(t), \text{RSPM}\,(t-1), \text{RSPM}\,(t-2)) \quad \text{for winter/summer/monsoon/postmonsoon} \tag{6.9}$$

Similarly, temporal seasonal 2 days ahead forecasting models can be written as

$$\text{SO}_2(t+2) = f(\text{SO}_2(t), \text{SO}_2(t-1), \text{SO}_2(t-2)) \quad \text{for winter/summer/monsoon/postmonsoon} \tag{6.10}$$

$$\text{NO}_x\ (t+2) = f(\text{NO}_x\ (t), \text{NO}_x\ (t-1), \text{NO}_x\ (t-2), \text{NO}_x\ (t-3)) \quad \text{for winter/summer/monsoon/postmonsoon} \tag{6.11}$$

$$\text{RSPM}\,(t+2) = f(\text{RSPM}\,(t), \text{RSPM}\,(t-1), \text{RSPM}\,(t-2)) \quad \text{for winter/summer/monsoon/postmonsoon} \tag{6.12}$$

One day and 2 days ahead, pollutants are predicted using ANN and GP with control parameters listed in Tables 6.1 and 6.2.

Table 6.1: Control parameters used in ANN.

S. no.	Items	Criteria used in this chapter
1	**Network architecture**	Input neurons = 3, 4, and 3 for SO_2, NO_x, and RSPM Output neurons = 1 variable for each model Hidden neurons = smallest number of neuron which yields minimum prediction error on the validation data set
2	**Neuron activation function**	Input neuron = identity function Output neuron = identity function Hidden neuron = hyperbolic tangent function logsig and purelin for all the models
3	**Learning parameters**	The learning parameters converge to the network configuration and give best performance on validation data set with least epochs
4	**Criteria for initialization of the initial weights**	Network weights are uniformly distributed in the range of −1 to 1
5	**Training algorithm**	Levenberg Marquardt
6	**Stopping criteria**	Performance goal/epochs
7	**Performance indicator**	***R*, RMSE, *d*, MBE**

Table 6.2: Control parameters used in GP.

S. no.	Parameters used	Value
1	Maximum initial tree size	15–20
2	Maximum tree size	45–50
3	Population size	50–1,000
4	No. of children produced	100–3,000
5	Mutation	0.05
6	Crossover rate	0.4–1.00
7	Objective type	R^2, RMSE

6.6 Results and discussion

We have demonstrated the application of GP, which is one of the evolutionary algorithms, a subset of machine learning to forecast the air quality parameters a few time steps in advance. One day and two days ahead, temporal annual and temporal seasonal models are developed and evaluated using ANN and GP. Owing to the numerous advantages, ANN has been commonly used for air quality forecasting [30]. It has the capability of adaptive learning self-organization and handling of the nonlinear systems [20]. We have used ANN as a benchmarking tool in this work. GP is relatively a new approach that has been used for air quality forecasting of Pune city. ANN and GP models have been developed and tested for unseen inputs, and the qualitative and quantitative performance is assessed by the correlation coefficient (r), root mean squared error (RMSE), d-statistics (d), and mean bias error (MBE).

6.6.1 Annual models

Temporal one day and two days ahead forecasting of air quality parameters has been carried out using ANN and GP.

6.6.1.1 Forecasting of SO_2

Sulfur dioxide is an air pollutant that is responsible for acid rain, haze, and many health-related problems. It is produced when the coal is burnt to generate electricity. As per CPCB report, 2010, the chief source of SO_2 in Pune city is industry

(73.53%) followed by mobile sources (12.70%) and brick kilns (6%). Industrial area in Pune has a very high fuel consumption. It is the largest forging industry in Asia. Observed concentrations of SO_2 are slightly higher than the permissible limits. Both the soft computing approaches, namely, ANN and GP worked reasonably well as evident from the performance indicators, but the performance of GP is slightly better than ANN for the present case study (refer Table 6.3). In the process of predictions using ANN, the structure is defined initially, and the learning algorithm finds the weights. In GP, functions are defined, which always results in an optimal solution. One day ahead prediction results are better than two days ahead as it can be observed from Fig. 6.3 and Fig. 6.4 as it can be seen with increased *r*.

Correlation coefficient of SO_2 $(t+1)$ for GP is 0.65 and for SO_2 $(t+2)$ is 0.6 and *d*-statistic of SO_2 $(t+1)$ for GP 0.8 and for SO_2 $(t+2)$ for GP is 0.75. The peak observed in testing for one day ahead prediction was 50 μg/cum and predicted by the ANN model is 44.35 μg/cum. The peak predicted by GP is 45.60 μg/cum. The observed peak for two days ahead forecasting was 46 μg/cum. Peak prediction by ANN is 33.81 μg/cum, and by GP it was 38.14 μg/cum.

6.6.1.2 Forecasting of NO_x

NO_x is produced in the air from the reaction of nitrogen and oxygen gases, especially at high temperatures, during combustion. The amount of nitrogen oxides emitted into the atmosphere as air pollutant can be significant, especially in the areas of high motor vehicle traffic in Pune city. Increased levels of nitrogen oxides can cause damage to the human respiratory tract and increase the risk of respiratory infections and asthma. Chronic lung diseases can be a result of long-term exposure to high levels of nitrogen dioxides. According to the CPCB report 2010, the significant contribution of 95% of NO_x in Pune is from vehicles. The industrial contribution is limited to 2% as industrial area is confined within the city limits. Remaining 3% of NO_x is produced from domestic and commercial fuel burning for cooking. Modeling and forecasting of NO_x can help to warn the residents, especially sensitive groups. It was observed that the concentrations of NO_x exceeded the annual standards during study period.

One day and two days ahead NO_x forecasting models have been developed and evaluated using ANN and GP. Alike SO_2, it can be seen that GP results are better than ANN and one day ahead predictions are superior than two days ahead (Tables 6.3 and 6.4 and also from Fig. 6.5 and Fig. 6.6). It can be seen that NO_x forecasts are more accurate than SO_2 forecasts. As far as errors are concerned, RMSE values are higher than as compared to SO_2 models, and negative MBE is observed for NO_x models indicating that the models are under predicting.

The peak observed in testing for one day ahead prediction was 115 μg/cum and predicted by the ANN model is 53.64 μg/cum. The peak predicted by GP is

Table 6.3: One day ahead temporal annual model results.

Performance evaluator	Models	Pollutant SO_2	NO_x	RSPM
r	ANN	0.6673	0.8110	0.8374
	GP	0.6776	0.8258	0.8434
RMSE	ANN	6.1876	11.6846	32.4299
	GP	6.0417	12.2770	31.6747
d	ANN	0.7989	0.8863	0.8956
	GP	0.7995	0.8943	0.9088
MBE	ANN	0.5077	0.2663	−2.1348
	GP	0.2041	−0.0391	−0.5727

Table 6.4: Two days ahead temporal annual model results.

Performance evaluator	Models	Pollutant SO_2	NO_x	RSPM
r	ANN	0.6523	0.8109	0.7006
	GP	0.6487	0.7752	0.7864
RMSE	ANN	6.2901	11.6856	42.3596
	GP	6.2298	12.6171	36.4318
d	ANN	0.7880	0.8862	0.8173
	GP	0.7585	0.8591	0.8713
MBE	ANN	0.4669	−0.2707	−2.1363
	GP	0.3714	−0.1105	−1.2505

123.45 μg/cum. Similarly, the observed peak for two days ahead forecasting was 155 μg/cum, and peak predicted by ANN and GP is 80.26 and 83.83 μg/cum, respectively, which indicates that GP predicts the peak closer to the observed values.

6.6.1.3 Forecasting of RSPM

Particulate matters with an aerodynamic diameter less than or equal to 10 μm are classified as respirable suspended particulates (RSP). Particulates are produced from combustion processes, vehicles, and industrial sources. They are microscopically small and can enter the lungs and penetrate deeply into the respiratory system and cause difficulty in breathing as well as respiratory and cardiovascular effects such as bronchitis and asthma. Sensitive groups, including pregnant woman, children, and older people, are at high risk. A warmer climate increases the risk of illnesses and death from extreme heat and poor air quality. Pune city has a hot and semiarid climate, which further increases the risk [32]. Vehicular emission is a major source of particulate matter in Pune and the highest recorded concentration of RSPM is 680 μg/cum during the study period. RSPM is the pollutant responsible for deciding the air quality. Modeling and forecasting of RSPM are essential to calculate the air quality index, which can be easily understood by the community.

From the observation table 6.3 and 6.4 and also from Fig. 6.7 and Fig. 6.8, it can be seen that similar to SO_2 and NO_x results, one day ahead forecasts of RSPM are better than two days ahead, and the GP models performed better than ANN. The fact that air pollution is a time-dependent phenomenon is verified from the results. There is a strong correlation between previous concentrations of RSPM with the forecasted values, which is evident from *r* values indicating a stronger correlation. "*r*" is a linear measure, and the air quality model evaluations based on only "*r*" mostly fail due to the presence of lag between source emission quantity and the ambient pollutant concentration. Temporal models fail to take into account the lag effect as the meteorological parameters are absent as inputs. The lag is observed due to inversion condition, which indicates the accumulation of pollutants in the ambient environment during odd hours of the day when there are no source emissions recorded [13]. In such conditions for air quality modeling, "*d*" statistics is the relevant evaluation criteria. "*d*" values of 0.90 and 0.87 are recorded for RSPM one day and two days ahead forecasting, which indicates that about 90% and 87% model results are error free. Negative MBE demonstrates that both the tools, ANN and GP, underpredict the phenomenon. Lower RMSE values (31.67 μg/cum for one day and 36.43 μg/cum for two days ahead forecasting) in the data values ranging between 17 and 680 μg/cum indicate the satisfactory performance of GP.

Alike SO_2 and NO_x models, peak predicted by GP for RSPM one day ahead model is closer to the actual value (observed peak is 291 μg/cum and the peak predicted by ANN is 242.55 μg/cum and by GP is 244.12 μg/cum), whereas the observed testing peak for two days ahead prediction was 680, but the peak predicted by ANN is 145.36 μg/cum and by GP is 153.28 μg/cum. This could be due to the fact that RSPM concertation of 680 μg/cum was observed only once during Diwali (festival) day.

6.6.2 Seasonal models

Seasonality is one of the essential factors affecting air quality. Variations in the pollutant concentrations are observed in different seasons due to the interaction of the various meteorological parameters with pollutant concentrations [27]. Seasonal models are formulated and evaluated using ANN and GP to get the complete picture of the air quality of Pune.

In Pune city, winter is observed during January and February. Air quality gets worsened during this time due to inversion condition. In winter time, a temperature inversion takes place when a layer of warmer air traps cold air near the ground. During inversion, the air becomes stagnant and pollutants are caught close to the land, which may result in the formation of smog.

Summer starts at the beginning of March and ends at the end of May. As compared to winter, summer heat prevents the inversion, which improves the air quality slightly. Summer always seems to drive out the dense clouds of pollution that suffocate many Indian citizens. Air pollutant concentrations are observed to be higher during the summer time because the heat and sunlight primarily react with the air along with all the chemical compounds in it. When these compounds react with the nitrogen oxide emissions present in the air, it results in smog. According to medical studies, air pollution in summer increases the risk of heart attack [31].

June to September is declared as monsoon months in Pune [26]. Rainfall is a cleansing mechanism that results in washing away of the pollutants during monsoon [26]. Rain makes the ground wet and mixing of the dust into the air again becomes difficult resulting in lesser pollutant concentration. Ozone formation rate is also lower in monsoon as the rain clouds obstruct the correct amount of temperature and sunlight required for ozone formation.

October and November are the two months of post monsoon (or retreating monsoon) in which the wind becomes weaker. Skies are clear, but humidity in the atmosphere increases. Diwali festival happens to be in this season. The air quality in this season decreases due to firecrackers. This condition increases the chance of pollution episodes.

All four seasons have different characteristics and from the observation tables, it can be seen that both the tools (ANN and GP) worked well, but GP results are better than ANN.

From the observation Tables 6.5–6.7, it can be seen that one day ahead predictions are superior to two days ahead forecasts. Results indicate that SO_2 models have shown better performance in summer and post-monsoon season. NO_x models have performed well in monsoon and post-monsoon and all RSPM models performed very well in all the seasons except for winter. The probable reason could be in winter there are slight movements of the air masses resulting in stagnation of the pollutants.

It was observed that maximum concentrations of SO_2, NO_x, and RSPM were recorded in winter, whereas the minimum concentrations have been recorded in

Table 6.5: Seasonal forecasting of SO_2.

Pollutant	Time of prediction	Season	Tool	Performance evaluator			
				r	RMSE	*d*	MBE
SO_2	1 day ahead	Winter	ANN	0.4734	8.4711	0.6688	0.7368
			GP	0.5577	7.5759	0.6963	0.3866
		Summer	ANN	0.6544	4.8672	0.7408	−0.0325
			GP	0.6546	4.7015	0.8267	0.3179
		Monsoon	ANN	0.5692	5.1757	0.7408	−0.0325
			GP	0.7067	4.3710	0.8267	0.8267
		Post monsoon	ANN	0.6338	7.8940	0.7865	0.6965
			GP	0.7066	6.8280	0.8219	0.3703
	2 days ahead	Winter	ANN	0.3024	9.0665	0.5142	1.0553
			GP	0.3588	8.6694	0.5503	−0.0517
		Summer	ANN	0.6022	5.4609	0.6044	2.0816
			GP	0.5914	5.0841	0.7040	0.9720
		Monsoon	ANN	0.5556	5.3268	0.7079	0.0609
			GP	0.5734	5.0645	0.7919	0.1971
		Post monsoon	ANN	0.4560	8.7618	0.6392	0.4464
			GP	0.4582	8.6370	0.6161	0.4400

monsoon as well as winter. Heavy fog during early mornings of winter supports the highest concentrations during winter. Rainfall was observed during study time in winter for few days resulting in the minimum pollutant concentrations. Rainfall acts as a cleansing agent and results in washing away of the pollutants in monsoon. From the meteorological records, it can be seen that the quantity of rainfall was sufficient to reduce the concentrations of the contaminants in monsoon.

6.7 Modeling options for situations when there are constraints on the data availability

Quality and consistency of the data decides the model prediction accuracy. Availability of continuous and accurate meteorological and pollutant concentration

Table 6.6: Seasonal forecasting of NO_x.

Pollutant	Time of prediction	Season	Tool	Performance evaluator			
				r	RMSE	*d*	MBE
NO_x	1 day ahead	Winter	ANN	0.6495	17.2756	0.7647	−1.1737
			GP	0.6846	16.4143	0.7849	−1.4931
		Summer	ANN	0.6758	19.3564	0.8681	−0.4552
			GP	0.6920	18.7333	0.8810	−0.1921
		Monsoon	ANN	0.7221	8.1978	0.8163	0.5864
			GP	0.7784	7.4318	0.8687	0.3330
		Post monsoon	ANN	0.7248	13.1202	0.8074	−1.7946
			GP	0.8080	11.0645	0.8848	0.0178
	2 days ahead	Winter	ANN	0.5625	18.6040	0.6885	−1.5606
			GP	0.6511	17.0322	0.7722	−0.6801
		Summer	ANN	0.5860	8.9691	0.7328	1.1892
			GP	0.5946	8.7939	0.7301	0.7584
		Monsoon	ANN	0.6352	9.4085	0.7155	2.0252
			GP	0.6775	8.7317	0.7903	0.5660
		Post monsoon	ANN	0.6245	14.9824	0.7135	−1.4086
			GP	0.6343	14.4177	0.7526	−0.0676

data is a challenging task in developing countries. Adverse meteorological conditions, instrumental errors, and so on are also some of the reasons for nonavailability of continuous and accurate data of all the meteorological parameters and pollutant concentrations. Atmospheric inversion is a vital condition deciding the air quality, which must be taken into account while modeling the air quality parameters. Data of meteorological parameters such as wind speed, rainfall, solar radiation, wind direction, humidity, mixing height, visibility, and color cover should be included as model inputs. Processing time for data collection and dissemination is quite high; as a result, it becomes difficult to include meteorology as a model input as in the case of this work. In such situations, the models developed should be robust enough so that they can be useful for short-term prediction of criteria air pollutants. An attempt has been made to suggest various air quality modeling alternatives.

Table 6.7: Seasonal forecasting of RSPM.

Pollutant	Time of prediction	Season	Tool	Performance evaluator			
				r	RMSE	*d*	MBE
RSPM	1 day ahead	Winter	ANN	0.5244	30.8937	0.6563	9.4319
			GP	0.5553	28.8443	0.6839	3.1098
		Summer	ANN	0.7935	19.3564	0.8681	−0.4552
			GP	0.8850	18.7333	0.8810	−0.1921
		Monsoon	ANN	0.7806	19.3564	0.8681	−0.4552
			GP	0.7964	18.7333	0.8810	−0.1921
		Post monsoon	ANN	0.6675	34.4305	0.7947	−0.5861
			GP	0.6826	33.6118	0.7914	−0.3259
	2 days ahead	Winter	ANN	0.3941	34.4683	0.8810	11.6724
			GP	0.3801	32.5242	0.8824	6.3909
		Summer	ANN	0.7729	52.3487	0.7881	−9.3464
			GP	0.7926	47.1344	0.8596	−3.0583
		Monsoon	ANN	0.6942	22.2994	0.8072	−0.5772
			GP	0.7104	21.7756	0.8122	−0.4177
		Post monsoon	ANN	0.4932	81.7014	0.4789	−31.2984
			GP	0.5527	14.4177	0.6815	−0.02559

When meteorological parameters are available, the correlation of inputs (meteorological parameters) with the output (pollutant concentrations) can be calculated and data can be arranged in the descending order of their correlation with the output. Only three inputs (from the available inputs) with high correlation with the output shall be considered and called as top three causes. Those having low correlation with the output are called as bottom three causes. The modeling options can be presented as [29]

$$\text{Model type 1 – model with top three inputs} \tag{6.13}$$

$$\text{Model type 2 – model with bottom three inputs} \tag{6.14}$$

$$\text{Model type 3 – model with bottom three and top one input} \tag{6.15}$$

$$\text{Model type 4 – temporal models} \tag{6.16}$$

Of the various meteorological parameters listed above, the temperature difference is the crucial factor for atmospheric inversion and, at the same time, it is effortless to record and disseminate. Temperature model can be developed in which previous values of the temperature difference can be added on the basis of correlation along with previous pollutant concentrations in case of the temporal model, and such modified temporal model can be presented as

$$\text{Model type 5} - \text{temperaturemodels} \tag{6.17}$$

With these modeling options, criteria air pollutants can be predicted on a continuous basis.

6.8 Conclusion and future directions

We have presented the application of soft computing techniques to forecast air quality parameters one and two days in advance. Models are developed with GP, and the results are compared with established forecasting tool of ANN. Temporal annual and seasonal models are developed for all the three criteria air pollutants (SO_2, NO_x, and RSPM) and found that model results are matching with the recorded values. One day ahead forecasting results are better than two days ahead. GP proved to be better than ANN in terms of accuracy of forecast and found equally competent for peak predictions for this case study.

In air quality modeling, researcher always prefers a short-term and continuous model to broadcast the current status of the air quality. This information is useful to plan the activities and safeguard the health of sensitive groups. In developing countries like India, there is a problem of availability of continuous data. In such situations, various modeling options are proposed.

The major limitation of the study is the absence of meteorological parameters for model development. Model inputs were selected using a linear method of correlation for the air quality forecasting problem, which is actually a nonlinear phenomenon. There were certain assumptions made while developing forecasting models such as seasonal patterns were assumed to be uniform, meteorological processes were simplified, and linearity between the inputs was assumed which is seldom true. The studies in future could be directed toward the better understanding of the tool GP and combining it with other air quality forecasting approaches. Comprehensive models can be prepared by considering the vertical and spatial distribution of the pollutants along with meteorological parameters.

References

[1] Ahmad K., Khare M., & Chaudhry K.K. (2005) Wind tunnel simulation studies on dispersion at urban street canyons and intersections –a review, Journal of Wind Engineering and Industrial Aerodynamics, 93, 697–717.

[2] Babovic V. & Keijzer M. (2000) Genetic programming as a model induction engine, Journal of Hydroinformatics, 2, 35–60.

[3] Brameier M. & Banzhaf W. (2001) A Comparison of linear genetic programming and neural networks in medical data mining, IEEE Transactions of Evolutionary Computations, 5(1), 17–26.

[4] Dixit P. & Londhe S.N. (2016) Prediction of Extreme wave heights using neuro wavelet techniques, Applied Ocean Research, 58, 241–252.

[5] Francone F.D. & Deschaine L.M. (2004) Extending the boundaries of design optimisation by integrating fast optimization techniques with machine code based linear genetic programming, Information Sciences, 161(3–4), 99–120.

[6] Gardner M.W. & Dorling S.R. (1998) Artificial neural networks (The Multilayer Perceptron)- A review of applications in the atmospheric sciences, Atmospheric Environment, 2627–2636.

[7] Gandomi A.H., Alavi A.H., & Yun G.J. (2011(b)) Nonlinear modelling of shear strength of SFRC beams using linear genetic programming, Structural Engineering Mechanics Techno press, 38(1), 1–25.

[8] Guven A. (2009) Linear Genetic Programming for time series modelling of daily flow rate, Journal of Earth System Sciences, 118(2), 137–146.

[9] Jakeman A.J., Simpson R.W., & Taylor J.A. (1988) Modeling distributions of air pollutant concentrations – III, The hybrid deterministic-statistical distribution approach. Atmospheric Environment, 22(1), 163–174.

[10] Londhe S.N. & Shah S. (2017) A Novel approach for knowledge extraction from artificial neural networks, ISH Journal of Hydraulic Engineering.

[11] Bai L., Wang J., Xuejio M., & Haiyan L. (2018) Air pollution forecasts: An overview, International Journal of Environmental Research and Public Health, 15(780), 01–44.

[12] Mishra D. & Goyal P. (2016) Neuro fuzzy approach to forecast NO_2 Pollutants addressed to air quality dispersion model over Delhi, India, Aerosol and Air Quality Research, 16, 166–174.

[13] Nagendra S.M.S. & Khare M. (2006) Artificial neural network approach for modelling nitrogen dioxide dispersion from vehicular exhaust emissions, Ecological Modelling, 190, 99–115.

[14] Niska H., Hiltunen T., Karppinen A., Ruuskanen J., & Kolehmainen M. (2004) Evolving the neural network model for forecasting air pollution time series, Engineering Applications of Artificial Intelligence, 17(2), 159–167.

[15] The ASCE task Committee. (2000) Artificial neural networks in hydrology. I: Preliminary concepts, Journal of Hydrologic Engineering, 115–123.

[16] Tikhe S.S., Khare K.C., & Londhe S.N. (2015) Multicity seasonal air quality index forecasting using soft computing techniques, Advances in Environmental Research, 4(02), 83–104.

[17] Yadav A. & Sharma D. Enhancement of performance 32 bit RISC processor using Genetic Algorithms, IJRECE, 7(3), 449–454.

[18] Lin Y., Zhao L., Haiyan L., & Sun Y. (2018) Air quality forecasting based on cloud model granulation, EURASIP Journal on Wireless Communications and Networking, 1–10.

[19] Zhang Y., Bocquet M., Mallet V., Seigneur C., & Baklanov A. (2012) Real-time air quality forecasting, part I: History, techniques, and current status, Atmospheric Environment, 60, 632–655.

Books

[20] Bose N.K. & Liang P. Neural network fundamentals with graphs, algorithms and applications. Tata McGraw Hill Publishing Company Limited, New Delhi, India, 1996.
[21] Cheremisinoff N.P. & Paul N. Cheremisinoff. Pumps/Compressors/Fans Pocket Handbook Lancaster PA. Technomic Publishing Company, 1989.
[22] Giorgio F. & Piero F. Mathematical models for planning and controlling air quality. Proceedings of IIASA workshop 17, 1996.
[23] Khare M. & Nagendra S.M.S. Artificial neural networks in vehicular pollution modelling. Springer, New York, 2007.
[24] Koza J.R. Genetic programming. MIT Press, 1992.
[25] Londhe S.N. & Dixit P. Genetic programming: A novel computing approach in modelling water flows. 2012.
[26] Londhe S.N. & Panchang V. (2018) ANN techniques: A survey of coastal applications, Advances in Coastal Hydraulics.
[27] Murthy P.B. Environmental Meteorology. I.K. International Publishing House Pvt. Ltd., New Delhi, 2009.
[28] Rumelhart D.E. & James M. Parallel Distributed Processing: Explorations in the Microstructure of Cognition. MIT Press, Cambridge, 1986.
[29] Tikhe Shruti K.K.C. & Londhe S.N. Forecasting of Air Quality Parameters using Soft Computing Techniques: Ph. D. thesis. Savitribai Phule Pune University, Pune, 2016.
[30] Weber E. Air pollution, assessment methodology and modeling. Plenum, New York, 1982.

Websites

[31] Air pollution in summer increases heart attack risks, 2019. (Accessed June,23, 2019 at https://www.iqair.com/intl/node/811.)
[32] Air quality monitoring and emission source opportionment study for Pune, 2010. (Accessed June, 12, 2019 at http://mpcb.gov.in/ereports/pdf/pune_report_cpcb.pdf.)
[33] Ambient air quality monitoring network in Maharashtra, 2016. (Accessed January, 15, 2016, at http://mpcb.gov.in/.)
[34] How air pollution is destroying our health, 2020. (Accessed July 30, 2019 at https://www.who.int/air-pollution/news-and-events/how-air-pollution-is-destroying-our-health.)
[35] India map, 2019 (Accessed June 23, 2019 at https://www.mapsofindia.com/.)
[36] Overview of The clean air act and air pollution, 2016. (Accessed January, 10, 2016 at https://www.epa.gov/clean-air-act-overview.)
[37] Pune air poses health threat to citizens, 2015. (Accessed December 4, 2015, at https://www.dnaindia.com/pune/report-pune-air-poses-health-threat-to-citizens-2141315.)

Palash Dutta, Bornali Saikia

7 Arithmetic operations on generalized semielliptic intuitionistic fuzzy numbers and their application in multicriteria decision making

Abstract: This chapter attempts to initiate a novel generalized semielliptic intuitionistic fuzzy number (GSEIFN) along with basic arithmetic operations on GSEIFNs. Furthermore, an advanced ranking method for GSEIFN is proposed. In the end, a multicriteria decision-making problem has been carried out by using the proposed GSEIFN and the ranking approach to exhibit the legality and applicability.

7.1 Introduction

Multicriteria decision making (MCDM) is a common activity in our daily life, such as choosing a car, selecting a candidate in an interview for the post of a teacher, and choosing or buying the best book or the other things from online shopping apps. It is an action or a procedure of choosing the best option from the accessible options through multiple criteria. In making decision, due to the presence of conflicting criteria or vague information, it is impossible to make proper decision by using a traditional set theory. To deal with such uncertain or vague data, Zadeh [1] introduced the fuzzy set theory (FST), which deals with those elements that partially belonged to the set. Later, Atanassov [2] extended FST, which is known as intuitionistic fuzzy set (IFS), which deals with both elements that are partially belonged to the set and not belonged to the set. IFS deals with such situations in an enhanced manner in comparison to FST. Due to growing complexities in making decisions, several intuitionistic fuzzy numbers (IFNs) have been developed [3–5]. More often, for simplicity, triangular or trapezoidal IFNs (TrIFNs) are employed to solve decision-making problems. However, in some real-world situations, the interactions between available system factors are found to be more complicated and in such situations triangular or trapezoidal types of IFNs fail to provide proper result. To make a proper decision, it is more essential to reconstruct an appropriate IFN. To overcome such difficulties, generalized semielliptic (GSE) type of IFN may be developed by making more suitable to handle such complex environments.

In MCDM problems, ranking of IFNs plays an efficient role. In literature, a few numbers of ranking approaches are encountered based on value and ambiguity [6], centroid point [7], centroid [3], magnitude [8], area [9], score and accuracy function

Palash Dutta, Bornali Saikia, Department of Mathematics, Dibrugarh University, Dibrugarh, India

https://doi.org/10.1515/9783110628616-007

[10], possibility mean [11], mean value [12], and improved possibility degree [13]. The approaches [3, 4, 6, 14–24] focused on the value as well as the ambiguity-based ranking method of IFNs and applied in MCDM problems. The approaches in [7] and [3] dealt with the centroid-based ranking method, the approach in [21] presented a ranking technique for IFNs by using the distance of each IFN from the fuzzy origin, and the approach in [9] developed a ranking method based on (α, β)-cuts and area of IFNs. Notwithstanding having some primacy of ranking approaches, these approaches have some deficiencies and do not provide a proper result, which produces predicament for judgment architects in any decision-making problem. To overcome such limitations, a new ranking approach has been developed based on the consummated exponential mean. Furthermore, a comparative analysis of the present ranking approach and other available techniques has been presented to exhibit the validity of the same. In this chapter, an effort has been made to introduce generalized semielliptic IFN (GSEIFN) and perform its fundamental operations of GSEIFNs. Finally, an MCDM problem is being solved to demonstrate the existence and significance of proposed IFNs and proposed ranking approach.

7.2 Preliminaries

Definition 7.1 ([2]). An IFS B defined on the universal set X is denoted by $B = \{(x, \mu_B(x), \nu_B(x)), x \in X\}$. μ_B:$X \to [0,1]$ and ν_B:$X \to [0,1]$ are respective membership function (MF) and nonmembership function (NMF) such that $0 \le \mu_B(x) + \nu_B(x) \le 1$. The amount $\pi_B(x) = 1 - \mu_B(x) - \nu_B(x)$ is called the degree of hesitancy.

Definition 7.2 ([6]). An α-cut of B is defined as

$$^{\alpha}B = \{x | \mu_B(x) \ge \alpha\}$$

For IFN α-cut produces a closed interval, say $^{\alpha}B = [L_B(\alpha), R_B(\alpha)]$.

Definition 7.3 ([6]). An β-cut of B is defined as

$$^{\beta}B = \{x | \nu_B(x) \le \beta\}$$

For an IFN the α-cut produces a closed interval, say $^{\beta}B = [L_B(\beta), R_B(\beta)]$.

Definition 7.4 ([6]). An (α, β)-cut of B is defined as

$$^{(\alpha,\beta)}B = \{x | \mu_B(x) \ge \alpha, \nu_B(x) \le \beta\}$$

such that $0 \le \alpha \le 1$, $0 \le \beta \le 1$, and $0 \le \alpha + \beta \le 1$.

Definition 7.5 ([18]). The values of IFN B of MF and NMF are denoted by $V_\mu(B)$ and $V_v(B)$, respectively, as

$$V_\mu(B) = \int_0^w [L_B(\alpha) + R_B(\alpha)]\alpha d\alpha$$

$$V_v(B) = \int_\eta^1 [L_B(\beta) + R_B(\beta)](1-\beta) d\beta$$

Definition 7.6 ([18]). The ambiguities of IFN B of MF and NMF are denoted by $A_\mu(B)$ and $A_v(B)$, respectively, as

$$A_\mu(B) = \int_0^w [L_B(\alpha) - R_B(\alpha)]\alpha d\alpha$$

$$A_v(B) = \int_\eta^1 [L_B(\beta) - R_B(\beta)](1-\beta) d\beta$$

Definition 7.7. The mean of IFN B are denoted $M_\mu(B)$ and $M_v(B)$, respectively, as

$$M_\mu(B) = \frac{1}{2}\int_0^w [L_B(\alpha) + R_B(\alpha)] d\alpha$$

$$M_v(B) = \frac{1}{2}\int_\eta^1 [L_B(\beta) + R_B(\beta)] d\beta$$

Definition 7.8 ([5]). The general expression of an ellipse centered at (p,q) is

$$\frac{(x-p)^2}{h^2} + \frac{(y-q)^2}{k^2} = 1$$

Assuming that $q = 0$ and $k = 1$ to construct a normal semielliptic IFN (NSEIFN) P, we have

$$\frac{(x-p)^2}{h^2} + y^2 = 1$$

Then the needed MF is

$$\mu_P(x) = \sqrt{1 - \frac{(x-p)^2}{h^2}}, p-h \le x \le p+h$$

In the same way, the NMF can also be constructed as

$$v_P(x) = 1 - \sqrt{1 - \frac{(x-p)^2}{h_1{}^2}}, p-h_1 \le x \le p+h_1$$

where $h \le h_1$ and p is the core of the NSEIFN P, while h and h_1 indicate support/domain MF and NMF, respectively.

Definition 7.9. The GSEIFN $P = S_E(p, h_1, h_2; w_1, \eta_1)$ on universal set of discourse X is defined by the MF and NMF, respectively, as

$$\mu_P(x) = w_1\sqrt{1 - \frac{(x-p)^2}{h_1^2}}, p - h_1 \le x \le p + h_1$$

$$\nu_P(x) = 1 - (1-\eta_1)\sqrt{1 - \frac{(x-p)^2}{h_2^2}}, p - h_2 \le x \le p + h_2$$

Here $h_1 \le h_2$.

Definition 7.10. Consider $P = S_E(p, h_1, h_2; w_1, \eta_1)$ be a GSEIFN. Then (α, β)-cut of the GSEIFN is

$${}^{\alpha}P = \left[p - h_1\sqrt{1 - \left\{\frac{\alpha}{w_1}\right\}^2}, p + h_1\sqrt{1 - \left\{\frac{\alpha}{w_1}\right\}^2}\right]$$

$${}^{\beta}P = \left[p - h_2\sqrt{1 - \left\{\frac{1-\beta}{1-\eta_1}\right\}^2}, p + h_2\sqrt{1 - \left\{\frac{1-\beta}{1-\eta_1}\right\}^2}\right]$$

7.3 Elementary operations on GSEIFNs

Here, we use (α, β)-cut method to perform some arithmetic operations of GSEIFNs and also discuss some numerical examples. Suppose that $P = S_E(p, h_1, h_2; w_1, \eta_1)$ and $Q = S_E(q, k_1, k_2; w_2, \eta_2)$ are two GSEIFNs whose MF and NMFs are defined as follows:

$$\mu_P(x) = w_1\sqrt{1 - \frac{(x-p)^2}{h_1^2}}, p - h_1 \le x \le p + h_1$$

$$\nu_P(x) = 1 - (1-\eta_1)\sqrt{1 - \frac{(x-p)^2}{h_2^2}}, p - h_2 \le x \le p + h_2$$

Here $h_1 \le h_2$

$$\mu_Q(x) = w_2\sqrt{1 - \frac{(x-q)^2}{k_1^2}}, q - k_1 \le x \le q + k_1$$

$$\nu_Q(x) = 1 - (1-\eta_2)\sqrt{1 - \frac{(x-q)^2}{k_2^2}}, q - k_2 \le x \le q + k_2$$

Here $k_1 \le k_2$.

Now we find out that the (α, β)-cuts of P and Q are

$$^{\alpha}P = \left[p - h_1\sqrt{1 - \left\{\frac{\alpha}{w_1}\right\}^2}, p + h_1\sqrt{1 - \left\{\frac{\alpha}{w_1}\right\}^2} \right]$$

$$^{\beta}P = \left[p - h_2\sqrt{1 - \left\{\frac{1-\beta}{1-\eta_1}\right\}^2}, p + h_2\sqrt{1 - \left\{\frac{1-\beta}{1-\eta_1}\right\}^2} \right]$$

$$^{\alpha}Q = \left[q - k_1\sqrt{1 - \left\{\frac{\alpha}{w_2}\right\}^2}, q + k_1\sqrt{1 - \left\{\frac{\alpha}{w_2}\right\}^2} \right]$$

$$^{\beta}Q = \left[q - k_2\sqrt{1 - \left\{\frac{1-\beta}{1-\eta_2}\right\}^2}, q + k_2\sqrt{1 - \left\{\frac{1-\beta}{1-\eta_2}\right\}^2} \right]$$

Here we also use $w_1 = \min(w_1, w_2)$ and $\eta_1 = \max(\eta_1, \eta_2)$ to perform arithmetic operations.

7.3.1 Addition of GSEIFNs

For the evaluation of sum of GSEIFNs P and Q, it is required to sum the α-cuts and β-cuts of P and Q via interval operations and here we assume $w_1 = \min(w_1, w_2)$, $\eta_1 = \max(\eta_1, \eta_2)$:

$$^{\alpha}P + {}^{\alpha}Q = \left[(p+q) - (h_1+k_1)\sqrt{1 - \left\{\frac{\alpha}{w_1}\right\}^2}, (p+q) + (h_1+k_1)\sqrt{1 - \left\{\frac{\alpha}{w_1}\right\}^2} \right] \tag{7.1}$$

$$^{\beta}P + {}^{\beta}Q = \left[(p+q) - (h_2+k_2)\sqrt{1 - \left\{\frac{1-\beta}{1-\eta_1}\right\}^2}, (p+q) + (h_2+k_2)\sqrt{1 - \left\{\frac{1-\beta}{1-\eta_1}\right\}^2} \right] \tag{7.2}$$

Now for getting proper MF $\mu_{P+Q}(x)$ and NMF $\nu_{P+Q}(x)$, it is important to equate initial component and final component of eqs. (7.1) and (7.2) with x which give

$$x = (p+q) - (h_1+k_1)\sqrt{1 - \left\{\frac{\alpha}{w_1}\right\}^2}, x = (p+q) - (h_2+k_2)\sqrt{1 - \left\{\frac{1-\beta}{1-\eta_1}\right\}^2}$$

$$x=(p+q)+(h_1+k_1)\sqrt{1-\left\{\frac{\alpha}{w_1}\right\}^2},x=(p+q)+(h_2+k_2)\sqrt{1-\left\{\frac{1-\beta}{1-\eta_1}\right\}^2}$$

Next, it is needed to express α in terms of x by taking $\alpha \geq 0$, $\alpha \leq w_1$ and $\alpha \leq w_1$, $\alpha \geq 0$ in eq. (7.1) we have

$$\alpha=w_1\sqrt{1-\left\{\frac{x-(p+q)}{h_1+k_1}\right\}^2},\alpha=w_1\sqrt{1-\left\{\frac{x-(p+q)}{h_1+k_1}\right\}^2}$$

We get the domain of x,

$$x\in[(p-h_1)+(q-k_1),(p+q)],x\in[(p+q),(p+h_1)+(q+k_1)]$$

In the same way, β can be expressed in terms of x by taking $\beta \leq \eta_1$, $\beta \geq 0$ and $\beta \geq 0$, $\beta \leq \eta_1$ in eq. (7.2) we have

$$\beta=1-(1-\eta_1)\sqrt{1-\left\{\frac{x-(p+q)}{h_2+k_2}\right\}^2},\beta=1-(1-\eta_1)\sqrt{1-\left\{\frac{x-(p+q)}{h_2+k_2}\right\}^2}$$

We get the domain of x,

$$x\in[(p-h_2)+(q-k_2),(p+q)],x\in[(p+q),(p+h_2)+(q+k_2)]$$

Thus, the MF $\mu_{P+Q}(x)$ and NMF $\nu_{P+Q}(x)$ of $P+Q$ are

$$\mu_{(P+Q)}=w_1\sqrt{1-\left\{\frac{x-(p+q)}{h_1+k_1}\right\}^2},x\in[(p-h_1)+(q-k_1),(p+h_1)+(q+k_1)]$$

$$\nu_{(P+Q)}=1-(1-\eta_1)\sqrt{1-\left\{\frac{x-(p+q)}{h_2+k_2}\right\}^2},x\in[(p-h_2)+(q-k_2),(p+h_2)+(q+k_2)]$$

7.3.2 Subtraction of GSEIFNs

For evaluation of subtraction of GSEIFNs P and Q, here, it is also required to subtract the corresponding α-cuts and β-cuts of P and Q via interval operations and considering $w_1=\min(w_1,w_2)$, $\eta_1=\max(\eta_1,\eta_2)$, we have

$$^{\alpha}P-{}^{\alpha}Q=\left[(p-q)-(h_1+k_1)\sqrt{1-\left\{\frac{\alpha}{w_1}\right\}^2},(p-q)+(h_1+k_1)\sqrt{1-\left\{\frac{\alpha}{w_1}\right\}^2}\right] \tag{7.3}$$

$$^{\beta}P-{}^{\beta}Q=\left[(p-q)-(h_2+k_2)\sqrt{1-\left\{\frac{1-\beta}{1-\eta_1}\right\}^2},(p-q)+(h_2+k_2)\sqrt{1-\left\{\frac{1-\beta}{1-\eta_1}\right\}^2}\right] \tag{7.4}$$

Now, for getting proper MF $\mu_{P-Q}(x)$ and NMF $\nu_{P-Q}(x)$ of $P-Q$, it is essential to equate both initial and final components with x eqs. (7.3) and (7.4) which give

$$x=(p-q)-(h_1+k_1)\sqrt{1-\left\{\frac{\alpha}{w_1}\right\}^2}, x=(p-q)-(h_2+k_2)\sqrt{1-\left\{\frac{1-\beta}{1-\eta_1}\right\}^2}$$

$$x=(p-q)+(h_1+k_1)\sqrt{1-\left\{\frac{\alpha}{w_1}\right\}^2}, x=(p-q)+(h_2+k_2)\sqrt{1-\left\{\frac{1-\beta}{1-\eta_1}\right\}^2}$$

Next, it is further essential to express α in terms of x by considering $\alpha \geq 0$, $\alpha \leq w_1$ and $\alpha \leq w_1$, $\alpha \geq 0$ in eq. (7.3) we have

$$\alpha=w_1\sqrt{1-\left\{\frac{x-(p-q)}{h_1+k_1}\right\}^2}, \alpha=w_1\sqrt{1-\left\{\frac{x-(p-q)}{h_1+k_1}\right\}^2}$$

We get the domain of x,

$$x \in [(p-h_1)-(q+k_1),(p-q)], x \in [(p-q),(p+h_1)-(q-k_1)]$$

In the same fashion, β can be expressed in terms of x by taking $\beta \leq \eta_1$, $\beta \geq 0$ and $\beta \geq 0$, $\beta \leq \eta_1$ in eq. (7.4) we have

$$\beta=1-(1-\eta_1)\sqrt{1-\left\{\frac{x-(p-q)}{h_2+k_2}\right\}^2}, \beta=1-(1-\eta_1)\sqrt{1-\left\{\frac{x-(p-q)}{h_2+k_2}\right\}^2}$$

We get the domain of x,

$$x \in [(p-h_2)-(q+k_2),(p-q)], x \in [(p-q),(p+h_2)-(q-k_2)]$$

Thus, the MF $\mu_{P-Q}(x)$ and NMF $\nu_{P-Q}(x)$ of $P-Q$ are

$$\mu_{(P-Q)}=w_1\sqrt{1-\left\{\frac{x-(p-q)}{h_1+k_1}\right\}^2}, x \in [(p-h_1)-(q+k_1),(p+h_1)-(q-k_1)]$$

$$\nu_{(P-Q)}=1-(1-\eta_1)\sqrt{1-\left\{\frac{x-(p-q)}{h_2+k_2}\right\}^2}, x \in [(p-h_2)-(q+k_2),(p+h_2)-(q-k_2)]$$

7.3.3 Product of GSEIFNs

For evaluating the product of GSEIFNs P and Q, it is required to product the α-cuts and β-cuts of P and Q via interval operations:

$$\begin{aligned} {}^{\alpha}P^{\alpha}Q = & \left[\left(p - h_1\sqrt{1 - \left\{\frac{\alpha}{w_1}\right\}^2} \right) \left(q - k_1\sqrt{1 - \left\{\frac{\alpha}{w_2}\right\}^2} \right), \right. \\ & \left. \left(p + h_1\sqrt{1 - \left\{\frac{\alpha}{w_1}\right\}^2} \right) \left(q + k_1\sqrt{1 - \left\{\frac{\alpha}{w_2}\right\}^2} \right) \right] \end{aligned} \tag{7.5}$$

$$\begin{aligned} {}^{\beta}P^{\beta}Q = & \left[\left(p - h_2\sqrt{1 - \left\{\frac{1-\beta}{1-\eta_1}\right\}^2} \right) \left(q - k_1\sqrt{1 - \left\{\frac{1-\beta}{1-\eta_2}\right\}^2} \right), \right. \\ & \left. \left(p + h_1\sqrt{1 - \left\{\frac{1-\beta}{1-\eta_1}\right\}^2} \right) \left(q + k_1\sqrt{1 - \left\{\frac{1-\beta}{1-\eta_2}\right\}^2} \right) \right] \end{aligned} \tag{7.6}$$

To find MF$\mu_{PQ}(x)$ and NMF $\nu_{PQ}(x)$, here too it is needed to equate first and last components to x in eqs. (7.5) and (7.6), which gives

$$x = \left(p - h_1\sqrt{1 - \left\{\frac{\alpha}{w_1}\right\}^2} \right) \left(q - k_1\sqrt{1 - \left\{\frac{\alpha}{w_2}\right\}^2} \right)$$

$$x = \left(p + h_1\sqrt{1 - \left\{\frac{\alpha}{w_1}\right\}^2} \right) \left(q + k_1\sqrt{1 - \left\{\frac{\alpha}{w_2}\right\}^2} \right)$$

$$x = \left(p - h_2\sqrt{1 - \left\{\frac{1-\beta}{1-\eta_1}\right\}^2} \right) \left(q - k_1\sqrt{1 - \left\{\frac{1-\beta}{1-\eta_2}\right\}^2} \right)$$

$$x = \left(p + h_1\sqrt{1 - \left\{\frac{1-\beta}{1-\eta_1}\right\}^2} \right) \left(q + k_1\sqrt{1 - \left\{\frac{1-\beta}{1-\eta_2}\right\}^2} \right)$$

Putting $w_1 = \min(w_1, w_2)$ and expressing α in terms of x

$$h_1k_1\left(1 - \left\{\frac{\alpha}{w_1}\right\}^2\right) - (pk_1 + qh_1)\left(\sqrt{1 - \left\{\frac{\alpha}{w_1}\right\}^2}\right) + pq - x = 0$$

$$h_1k_1\left(1 - \left\{\frac{\alpha}{w_1}\right\}^2\right) + (pk_1 + qh_1)\left(\sqrt{1 - \left\{\frac{\alpha}{w_1}\right\}^2}\right) + pq - x = 0$$

Next, solving the expression by taking $1 - \left\{\frac{\alpha}{w_1}\right\}^2 = y^2$, we have,

$$y = \frac{-\{-(pk_1 + qh_1)\} \pm \sqrt{\{-(pk_1 + qh_1)\}^2 - 4h_1k_1(pq - x)}}{2h_1k_1}$$

$$\alpha = w_1\sqrt{1 - \left\{\frac{(pk_1 + qh_1) - \sqrt{(pk_1 + qh_1)^2 - 4h_1k_1(pq - x)}}{2h_1k_1}\right\}^2}$$

Similarly,

$$\alpha = w_1\sqrt{1 - \left\{\frac{(pk_1 + qh_1) - \sqrt{(pk_1 + qh_1)^2 - 4h_1k_1(pq - x)}}{2h_1k_1}\right\}^2}$$

Setting $\alpha \geq 0$, $\alpha \leq w_1$ and $\alpha \leq w_1$, $\alpha \geq 0$ in eq. (7.5), we have

$$x \in [(p - h_1)(q - k_1), pq], x \in [pq, (p + h_1)(q + k_1)]$$

Putting $\eta_1 = \max(\eta_1, \eta_2)$ and expressing β in terms of x

$$h_2k_2\left\{1 - \left\{\frac{1-\beta}{1-\eta_1}\right\}^2\right\} - (pk_2 + qh_2)\sqrt{1 - \left\{\frac{1-\beta}{1-\eta_1}\right\}^2} + pq - x = 0$$

$$h_2k_2\left\{1 - \left\{\frac{1-\beta}{1-\eta_1}\right\}^2\right\} + (pk_2 + qh_2)\sqrt{1 - \left\{\frac{1-\beta}{1-\eta_1}\right\}^2} + pq - x = 0$$

Again solving the expression by taking $1 - \left\{\frac{1-\beta}{1-\eta_1}\right\}^2 = q^2$, we have

$$q = \frac{-\{-(pk_2 + qh_2)\} \pm \sqrt{\{-(pk_2 + qh_2)\}^2 - 4h_2k_2(pq - x)}}{2h_2k_2}$$

$$\beta = 1 - (1 - \eta_1)\sqrt{1 - \left\{\frac{(pk_2 + qh_2) - \sqrt{(pk_2 + qh_2)^2 - 4h_2k_2(pq - x)}}{2h_2k_2}\right\}^2}$$

Similarly

$$\beta = 1 - (1 - \eta_1)\sqrt{1 - \left\{\frac{(pk_2 + qh_2) - \sqrt{(pk_2 + qh_2)^2 - 4h_2k_2(pq - x)}}{2h_2k_2}\right\}^2}$$

Setting $\beta \leq \eta_1$, $\beta \geq 0$ and $\beta \geq 0$, $\beta \leq \eta_1$ in eq. (7.6) we have

$$x \in [(p - h_2)(q - k_2), pq], x \in [pq, (p + h_2)(q + k_2)]$$

Thus, the MF $\mu_{PQ}(x)$ and NMF $\nu_{PQ}(x)$ of PQ are

$$\mu_{PQ}(x) = w_1\sqrt{1-\left\{\frac{(pk_1+qh_1)-\sqrt{(pk_1+qh_1)^2-4h_1k_1(pq-x)}}{2h_1k_1}\right\}^2}$$

$$(p-h_1)(q-k_1)\le x\le (p+h_1)(q+k_1)$$

$$\nu_{PQ}(x) = 1-(1-\eta_1)\sqrt{1-\left\{\frac{(pk_2+qh_2)-\sqrt{(pk_2+qh_2)^2-4h_2k_2(pq-x)}}{2h_2k_2}\right\}^2}$$

$$(p-h_2)(q-k_2)\le x\le (p+h_2)(q+k_2)$$

7.3.4 Division of GSEIFNs

For evaluation division of GSEIFNs P by Q, it is essential to divide the α-cuts by β-cuts of P and Q via interval operations:

$$\frac{{}^{\alpha}P}{{}^{\alpha}Q} = \left[\frac{p-h_1\sqrt{1-\left\{\frac{\alpha}{w_1}\right\}^2}}{q+k_1\sqrt{1-\left\{\frac{\alpha}{w_2}\right\}^2}}, \frac{p+h_1\sqrt{1-\left\{\frac{\alpha}{w_1}\right\}^2}}{q-k_1\sqrt{1-\left\{\frac{\alpha}{w_2}\right\}^2}}\right] \tag{7.7}$$

$$\frac{{}^{\beta}P}{{}^{\beta}Q} = \left[\frac{p-h_2\sqrt{1-\left\{\frac{1-\beta}{1-\eta_1}\right\}^2}}{q+k_2\sqrt{[1-\left\{\frac{1-\beta}{1-\eta_2}\right\}^2}}, \frac{p+h_2\sqrt{1-\left\{\frac{1-\beta}{1-\eta_1}\right\}^2}}{q-k_2\sqrt{1-\left\{\frac{1-\beta}{1-\eta_2}\right\}^2}}\right] \tag{7.8}$$

Now, for getting exact MF $\mu_{\frac{A}{B}}(x)$ and NMF $\nu_{\frac{A}{B}}(x)$ of P/Q, it is required to equate first and last components to x in eqs. (7.7) and (7.8), we have

$$x = \frac{p-h_1\sqrt{1-\left\{\frac{\alpha}{w_1}\right\}^2}}{q+k_1\sqrt{1-\left\{\frac{\alpha}{w_2}\right\}^2}}, x = \frac{p+h_1\sqrt{1-\left\{\frac{\alpha}{w_1}\right\}^2}}{q-k_1\sqrt{1-\left\{\frac{\alpha}{w_2}\right\}^2}}$$

$$x = \frac{p-h_2\sqrt{1-\left\{\frac{1-\beta}{1-\eta_1}\right\}^2}}{q+k_2\sqrt{1-\left\{\frac{1-\beta}{1-\eta_2}\right\}^2}}, x = \frac{p+h_2\sqrt{1-\left\{\frac{1-\beta}{1-\eta_1}\right\}^2}}{q-k_2\sqrt{1-\left\{\frac{1-\beta}{1-\eta_2}\right\}^2}}$$

Next, express α in terms of x by considering $w_1 = \min(w_1, w_2)$, $\alpha \ge 0$, $\alpha \le w_1$ and $\alpha \le w_1$, $\alpha \ge 0$ in eq. (7.7) we have

$$\alpha = w_1\sqrt{1-\left\{\frac{p-xq}{h_1+xk_1}\right\}^2}, \quad \alpha = w_1\sqrt{1-\left\{\frac{p-xq}{h_1+xk_1}\right\}^2}$$

We get the domain of x

$$x \in \left[\frac{p-h_1}{q+k_1}, \frac{p}{q}\right], \quad x \in \left[\frac{p}{q}, \frac{p+h_1}{q-k_1}\right]$$

In the same fashion, β can be expressed in terms of x by taking $\eta_1 = \max(\eta_1, \eta_2)$, $\beta \le \eta_1$, $\beta \ge 0$ and $\beta \ge 0$, $\beta \le \eta_1$ in eq. (7.8) we have

$$\beta = 1-(1-\eta_1)\sqrt{1-\left\{\frac{p-xq}{h_2+xk_2}\right\}^2}, \quad \beta = 1-(1-\eta_1)\sqrt{1-\left\{\frac{p-xq}{h_2+xk_2}\right\}^2}$$

We get the domain of x

$$x \in \left[\frac{p-h_2}{q+k_2}, \frac{p}{q}\right], \quad x \in \left[\frac{p}{q}, \frac{p+h_2}{q-k_2}\right]$$

Thus, the MF $\mu_{\frac{P}{Q}}(x)$ and NMF $\nu_{\frac{P}{Q}}(x)$ of $\frac{P}{Q}$ are

$$\mu_{\frac{P}{Q}}(x) = w_1\sqrt{1-\left\{\frac{p-xq}{h_1+xk_1}\right\}^2}, \quad \frac{p-h_1}{q+k_1} \le x \le \frac{p+h_1}{q-k_1}$$

$$\nu_{\frac{P}{Q}}(x) = 1-(1-\eta_1)\sqrt{1-\left\{\frac{p-xq}{h_2+xk_2}\right\}^2}, \quad \frac{p-h_2}{q+k_2} \le x \le \frac{p+h_2}{q-k_2}$$

It should be noted that fundamental operations on GSEIFNs produces GSEIFN.

7.3.5 Illustrative examples

Here, numerical examples of GSEIFNs are being discussed with the help of above arithmetic operations.

Suppose that $P = S_E(10, 4, 5; 0.2, 0.5)$ and $Q = S_E(8, 2, 3; 0.5, 0.5)$ are GSEIFNs and their respective MFs $\mu_P(x)$, $\mu_Q(x)$ and NMFs $\nu_P(x)$, $\nu_Q(x)$ are as follows:

$$\mu_P(x) = 0.2\sqrt{1-\left(\frac{x-10}{4}\right)^2}, \quad x \in [6, 14]$$

$$\nu_P(x) = 1-(1-0.5)\sqrt{1-\left(\frac{x-10}{5}\right)^2}, \quad x \in [5, 15]$$

$$\mu_Q(x) = 0.5\sqrt{1 - \left(\frac{x-8}{2}\right)^2}, \quad x \in [6, 10]$$

$$\nu_Q(x) = 1 - (1 - 0.5)\sqrt{1 - \left(\frac{x-8}{3}\right)^2}, \quad x \in [5, 11]$$

The (α, β)-cuts of P and Q are

$$^{\alpha}P = \left[10 - 4\sqrt{1 - \left\{\frac{\alpha}{0.2}\right\}^2}, 10 + 4\sqrt{1 - \left\{\frac{\alpha}{0.2}\right\}^2}\right]$$

$$^{\beta}P = \left[10 - 5\sqrt{1 - \left\{\frac{1-\beta}{1-0.5}\right\}^2}, 10 + 5\sqrt{1 - \left\{\frac{1-\beta}{1-0.5}\right\}^2}\right]$$

$$^{\alpha}Q = \left[8 - 2\sqrt{1 - \left\{\frac{\alpha}{0.5}\right\}^2}, 8 + 2\sqrt{1 - \left\{\frac{\alpha}{0.5}\right\}^2}\right]$$

$$^{\beta}Q = \left[8 - 3\sqrt{1 - \left\{\frac{1-\beta}{1-0.5}\right\}^2}, 8 + 3\sqrt{1 - \left\{\frac{1-\beta}{1-0.5}\right\}^2}\right]$$

7.3.5.1 Sum of GSEIFNs

Sum of GSEIFNs P and Q is given by $P + Q$ whose MF $\mu_{P+Q}(x)$ and NMF $\nu_{P+Q}(x)$ are

$$\mu_{P+Q}(x) = 0.2\sqrt{1 - \left(\frac{x-18}{6}\right)^2}, \quad x \in [12, 24]$$

$$\nu_{P+Q}(x) = 1 - 0.5\sqrt{1 - \left(\frac{x-18}{8}\right)^2}, \quad x \in [10, 26]$$

which shows this is again a GSEIFN.

7.3.5.2 Subtraction of GSEIFNs

Subtraction of GSEIFNs P and Q is given by $P - Q$ where MF $\mu_{P-Q}(x)$ and NMF $\nu_{P-Q}(x)$ are

$$\mu_{P-Q}(x) = 0.2\sqrt{1-\left(\frac{x-2}{6}\right)^2}, \quad x \in [-4, 8]$$

$$\nu_{P-Q}(x) = 1-0.5\sqrt{1-\left(\frac{x-2}{8}\right)^2}, \quad x \in [-6, 10]$$

which shows that subtraction of two GSEIFNs P and Q is again a GSEIFN.

7.3.5.3 Product of GSEIFNs

Product of GSEIFNs P and Q is given by PQ where MF $\mu_{PQ}(x)$ and NMF $\nu_{PQ}(x)$ are

$$\mu_{PQ}(x) = 0.2\sqrt{1-\left\{\frac{52-\sqrt{(52)^2-32(80-x)}}{16}\right\}^2},$$
$$36 \le x \le 140$$

$$\nu_{PQ}(x) = 1-0.5\sqrt{1-\left\{\frac{70-\sqrt{(70)^2-60(80-x)}}{30}\right\}^2},$$
$$25 \le x \le 165$$

which shows that multiplication of two GSEIFNs P and Q is again a GSEIFN.

7.3.5.4 Division of GSEIFNs

Division of GSEIFNs P and Q is given by $\frac{P}{Q}$ whose MF $\mu_{\frac{P}{Q}}(x)$ and NMF $\nu_{\frac{P}{Q}}(x)$ are

$$\mu_{\frac{P}{Q}}(x) = 0.2\sqrt{1-\left(\frac{10-8x}{4+2x}\right)^2}, \quad \frac{6}{10} \le x \le \frac{14}{6}$$

$$\nu_{\frac{P}{Q}}(x) = 1-0.5\sqrt{1-\left(\frac{10-8x}{5+3x}\right)^2}, \quad \frac{5}{11} \le x \le \frac{15}{5}$$

which shows that division of two GSEIFNs P and Q is again a GSEIFN.

7.4 Mean and value of GSEIFN

Here, an attempt has been made to study mean and value of the newly defined GSEIFN by taking (α,β)-cuts. Suppose that $P = S_E(p, h_1, h_2; w_1, \eta_1)$ is a GSEIFN and its MF as well as NMF are as presented in Section 7.2. The mean of GSEIFN is evaluated from Definition 7.7:

$$M_\mu(P) = \frac{1}{2}\int_0^{w_1} [L_P(\alpha) + R_P(\alpha)]d\alpha$$

$$= \frac{1}{2}\int_0^{w_1} \left[p - h_1\sqrt{1-\left\{\frac{\alpha}{w_1}\right\}^2} + p + h_1\sqrt{1-\left\{\frac{\alpha}{w_1}\right\}^2} \right] d\alpha$$

$$= pw_1$$

Similarly for NMF $M_\nu(P) = p(1-\eta_1)$.

The value of GSEIFN P is calculated as

$$V_\mu(P) = \int_0^{w_1} [L_P(\alpha) + R_P(\alpha)]\alpha d\alpha$$

$$= pw_1^2$$

Similarly for NMF $V_\nu(P) = p(1-\eta_1)^2$.

7.5 Comparing the proposed IFN with the other existing IFNs

In this section, we compare the introduced IFN with the other existing IFNs by using some numerical examples and existing ranking approaches. Mainly we compare GSEIFN with the normal trapezoidal (NTrIFN), normal triangular IFN (NTIFN), and NSEIFN with the help Rezvani [17] and Z. Xu's [25] ranking method as discussed below.

7.5.1 Ranking approach of Rezvani

Rezvani [17] proposed the ranking approach for TrIFNs based on the value. We can easily evaluate the rank of two different TrIFNs by using this method. By this approach, values for the TrIFN $P = \langle a_1, a_2, a_3, a_4; a_1', a_2, a_3, a'_4 \rangle$ are evaluated as

$$V_\mu(A) = \frac{a_1 + 2a_2 + 2a_3 + a_4}{6}$$

For two different TrIFNs A and B, rank can be evaluated by using the following procedure:

$$\text{If } V_\mu(A) > V_\mu(B) \text{ then } A > B$$

$$\text{If } V_\mu(A) \le V_\mu(B) \text{ then } A \le B$$

This approach gives the counterintuitive output in some situations. For example, for the NTrIFNs $A = <(4, 5.5, 6, 8);1, 0>$ and $B = <(3.5, 5, 7, 7.5);1, 0>$ values of A and B denoted by $V_\mu A$ and $V_\mu B$ are evaluated as

$$V_\mu(A) = \frac{4 + 2\times 5.5 + 2\times 6 + 8}{6} = 5.833$$

$$V_\mu(B) = \frac{3.5 + 2\times 5 + 2\times 7 + 7.5}{6} = 5.833$$

This gives the same output for non identical NTrIFNs A and B. It is problematic for the decision makers to identify the best one. To get rid of this situation, NTrIFN can be reconstructed in the form of GSEIFN as given below.

Suppose that $A = S_E(6, 2, 2; 0.5, 0.5)$ and $B = S_E(5.5, 2, 2; 0.1, 0.7)$ are two GSEIFNs whose MFs and NMFs are

$$\mu_{(A)(x)} = 0.5\sqrt{1 - \left\{\frac{x-6}{2}\right\}^2}, \quad x \in [4, 6]$$

$$\nu_{(B)(x)} = 1 - 0.5\sqrt{1 - \left\{\frac{x-6}{2}\right\}^2}, \quad x \in [4, 6]$$

and

$$\mu_{(B)(x)} = 0.1\sqrt{1 - \left\{\frac{x-5.5}{2}\right\}^2}, \quad x \in [3.5, 7.5]$$

$$\nu_{(B)(x)} = 1 - 0.3\sqrt{1 - \left\{\frac{x-5.5}{2}\right\}^2}, \quad x \in [3.5, 7.5]$$

Then α-cuts and β-cuts of A and B are

$${}^{\alpha}A = \left[6 - 2\sqrt{1 - \left\{\frac{\alpha}{0.5}\right\}^2}, 6 + 2\sqrt{1 - \left\{\frac{\alpha}{0.5}\right\}^2}\right]$$

$$^{\alpha}B=\left[5.5-2\sqrt{1-\left\{\frac{\alpha}{0.1}\right\}^2},5.5+2\sqrt{1-\left\{\frac{\alpha}{0.1}\right\}^2}\right]$$

$$^{\beta}A=\left[6-2\sqrt{1-\left\{\frac{1-\beta}{0.5}\right\}^2},6+2\sqrt{1-\left\{\frac{1-\beta}{0.5}\right\}^2}\right]$$

$$^{\beta}B=\left[5.5-2\sqrt{1-\left\{\frac{1-\beta}{0.3}\right\}^2},5.5+2\sqrt{1-\left\{\frac{1-\beta}{0.3}\right\}^2}\right]$$

Through Definition 7.5, the values for A and B can be measured as

$$V_\mu(A)=\int_0^{0.5}\left(6-2\sqrt{1-\left\{\frac{\alpha}{0.5}\right\}^2}+6+2\sqrt{1-\left\{\frac{\alpha}{0.5}\right\}^2}\right)\alpha d\alpha=1.5$$

$$V_\mu(B)=\int_0^{0.1}\left(5.5-2\sqrt{1-\left\{\frac{\alpha}{0.1}\right\}^2}+5.5+2\sqrt{1-\left\{\frac{\alpha}{0.1}\right\}^2}\right)\alpha d\alpha=0.055$$

which shows that $V_\mu(A)>V_\mu(B)$ and this indicates that $A>B$. It is witness that drawbacks of earlier method can be overcome.

7.5.2 Ranking approach of Xu

Using score and accuracy function, Xu [25] developed this ranking approach [25]. Suppose that $A=<a_1,a_2,a_3;w_1,\eta_1>$ and $B=<b_1,b_2,b_3;w_2,\eta_2>$ are two triangular IFNs. Then by Xu's ranking method, score functions denoted by $S(A),S(B)$ and accuracy functions denoted by $H(A),H(B)$ are as follows:

$$S(A)=w_1-\eta_1,S(B)=w_2-\eta_2,H(A)=w_1+\eta_1,H(B)=w_2+\eta_2$$

Then the rank of A and B can be determined by using the following procedure:

If $S(A)<S(B)$ then $A<B$

If $S(A)=S(B)$ then we check the value of accuracy function

If $H(A)>H(B)$ then $A>B$

If $H(A)\le H(B)$ then $A\le B$

For the two NTIFNs $A=<(4,5.5,6);1,0>$ and $B=<(3.5,5,7);1,0>$, using Xu's approach the score and accuracy values are obtained as

$$S(A)=1,S(B)=1,H(A)=1,H(B)=1$$

which provides the identical result.

Similarly, NSEIFNs $A = <(4,2,3);1,0>$ and $B = <(8,3,5);1,0>$ whose MFs and NMFs are

$$\mu_{(A)(x)} = \sqrt{1 - \left\{\frac{x-4}{2}\right\}^2}, x \in [2,6]$$

$$\nu_{(B)(x)} = 1 - \sqrt{1 - \left\{\frac{x-4}{3}\right\}^2}, x \in [1,7]$$

and

$$\mu_{(B)(x)} = \sqrt{1 - \left\{\frac{x-8}{3}\right\}^2}, x \in [5,11]$$

$$\nu_{(B)(x)} = 1 - \sqrt{1 - \left\{\frac{x-8}{5}\right\}^2}, x \in [3,13]$$

For these IFNs, value score and accuracy functions are

$$S(A) = 1, S(B) = 1, H(A) = 1, H(B) = 1$$

These IFNs also provide identical results. So it is difficult for the experts to distinguish the best alternative by using NSEIFN. So we reconstruct the IFN in the form of GSE type. Suppose that $A = S_E(4,2,3;0.1,0.05)$ and $B = S_E(8,3,5;0.2,0.03)$ whose MF and NMFs are

$$\mu_{(A)(x)} = 0.1\sqrt{1 - \left\{\frac{x-4}{2}\right\}^2}, x \in [2,6]$$

$$\nu_{(B)(x)} = 1 - 0.95\sqrt{1 - \left\{\frac{x-6}{3}\right\}^2}, x \in [3,9]$$

and

$$\mu_{(B)(x)} = 0.2\sqrt{1 - \left\{\frac{x-8}{3}\right\}^2}, x \in [5,11]$$

$$\nu_{(B)(x)} = 1 - 0.97\sqrt{1 - \left\{\frac{x-8}{5}\right\}^2}, x \in [3,13]$$

Then the values of score functions and accuracy functions are

$$S(A) = 0.05, S(B) = 0.17, H(A) = 0.15, H(B) = 0.23$$

which shows that $A < B$. From this comparative analysis, it can be opined that the present approach has the ability to overcome those drawbacks of NTIFNs and NSEIFNs.

7.6 Ranking approach of IFN based on consummated exponential mean

Here, a novel ranking approach of IFNs based on the consummated exponential mean has been presented.

Let B be an IFN whose (α, β)-cuts are considered as follows:

$$^{\alpha}B = [L_B(\alpha), R_B(\alpha)], \quad ^{\beta}B = [L_B(\beta), R_B(\beta)]$$

Now, evaluate the mean of B as

$$M_\mu(B) = \frac{1}{2}\int_0^w [L_B(\alpha) + R_B(\alpha)]d\alpha$$

$$M_\nu(B) = \frac{1}{2}\int_\eta^1 [L_B(\beta) + R_B(\beta)]d\beta$$

for normal $w = 1$ and $\eta = 0$.

Now the consummated exponential mean of B is given by $M(B)$ and is defined as

$$M(B) = e^{M_\mu(B)} + e^{M_\nu(B)}$$

Suppose that P and Q are two IFNs, then the ranking procedure is as follows:

$$\text{If } M(P) > M(Q) \text{ then } P > Q$$

$$\text{If } M(P) \leq M(Q) \text{ then } P \leq Q$$

Consummated exponential mean of generalized trapezoidal IFN (GTrIFN) $P = < [p_1, p_2, p_3, p_4; w_1], [p_1', p_2, p_3, p_4'; \eta_1] >$ is defined by

$$M(P) = e^{\left\{\frac{w_1}{4}(p_1 + p_2 + p_3 + p_4)\right\}} + e^{\left\{\frac{(1-\eta_1)}{4}(p_1' + p_2 + p_3 + p_4')\right\}}$$

Consummated exponential mean of generalized triangular IFN (GTIFN) $P < [p_1, p_2, p_4; w_1], [p_1', p_2, p_4'; \eta_1] >$ is defined by

$$M(P) = e^{\{\frac{w_1}{4}(p_1 + 2p_2 + p_4)\}} + e^{\{\frac{(1-\eta_1)}{4}(p_1' + 2p_2 + p_4')\}}$$

Consummated exponential mean of GSEIFN $P = (p, h_1, h_2; w_1, \eta_1)$ is defined by

$$M(P) = e^{pw_1} + e^{p(1-\eta_1)}$$

7.6.1 Numerical illustration

Here, a discussion has been made on some numerical examples with the help of proposed ranking approach. Suppose that $P = < (1, 2, 3, 4; 0.2), (0.5, 2, 3, 4.5; 0.4) >$

are two GTrIFNs. Then the consummated exponential mean of P and Q are evaluated as follows:

$$\begin{aligned} M(P) &= e^{\left\{\frac{0.2}{4}(1+2+3+4)\right\}} + e^{\left\{\frac{(1-0.4)}{4}(0.5+2+3+4)\right\}} \\ &= e^{0.5} + e^{1.5} \\ &= 6.130411 \end{aligned}$$

Similarly, $M(Q) = 3.799609$, which shows that $M(P) > M(Q)$, that is the rank of order $P > Q$.

Suppose that $P = <(1, 2, 3; 0.2), (0.5, 2, 3; 0.4)>$ are two GTIFNs. Then the consummated exponential mean of P and Q are evaluated as follows:

$$\begin{aligned} M(P) &= e^{\left\{\frac{0.2}{4}(1+2+3)\right\}} + e^{\left\{\frac{(1-0.4)}{4}(0.5+2+3)\right\}} \\ &= e^{0.4} + e^{1.125} \\ &= 4.572042 \end{aligned}$$

Similarly, $M(Q) = 2.645911$,

which shows that $M(P) > M(Q)$, that is, the rank of order $P > Q$.

Suppose that $P = S_E(5, 2, 3; 0.2, 0.3)$ and $P = S_E(4, 1, 2; 0.1, 0.3)$ are two GSEIFNs, then the consummated exponential mean of P and Q are

$$M(P) = 4.137349, \quad M(Q) = 3.717366$$

which shows that $M(P) > M(Q)$, that is, the rank of order $P > Q$.

7.6.2 Comparative study of the present approach with the available approaches

Here, we compare the present ranking approach with the other existing methods in literature as shown in Table 7.1.

From the above table it is observed that the ranking approaches developed in [6, 10, 15, 17, 26, 27] as well as in [9] produce the same result for nonidentical IFNs. To overcome these drawbacks and to get a proper result, we should develop a proper ranking approach. In this chapter, we introduced a consummated exponential mean-based ranking approach and from the comparative Table 7.1, it is encountered that the present ranking approach has the capability to overcome the limitations of the available methods.

Table 7.1: Comparative table.

IFNs	Available approaches	Present approach
$P = <(4, 5.5, 6); 1, 0>$, $Q = <(3.5, 5, 7); 1, 0>$	[25], $P = Q$	$P > Q$
$P = <(0.45, 0.63, 0.83); 0.73, 0.2>$, $Q = <(0.48,\ 0.64,\ 0.82);\ 0.72,\ 0.2>$	[10], $P = Q$	$P < Q$
$P = \langle(0, 0.25, 0.3); 1, 0\rangle$, $Q = \langle(0.1, 0.2, 0.4); 1, 0\rangle$	[15], $P = Q$	$P < Q$
$P = <(0, 0.25, 0.3); 1, 0>$, $Q = <(0.1, 0.2, 0.4); 1, 0>$	[6], $P = Q$	$P < Q$
$P = <(0.20, 0.35, 0.40, 0.55); 1, 0>$, $Q = <(0.10, 0.40, 0.45, 0.45); 1, 0>$	[26], $P = Q$	$P > Q$
$P = <(4, 5.5, 6, 8); 1, 0>$, $Q = <(3.5, 5, 7, 7.5); 1, 0>$	[17], $P = Q$	$P > Q$
$P = <(2.57, 2.73, 2.83); 0.73, 0.2>$, $Q = <(2.58, 2.74, 2.82); 0.72, 0.2>$	[27], $P = Q$	$P > Q$
$P = <(3.5, 5, 7, 7.5); 0.5, 0.5>$, $Q = <(4, 5.5, 6, 8); 0.7, 0.3>$	[7], $P > Q$	$P < Q$
$P = <(0.20, 0.35, 0.40, 0.55); 1, 0>$, $Q = <(0.10, 0.40, 0.45, 0.45); 1, 0>$	[28], $P < Q$	$P > Q$
$P = <(4, 4, 4); 1, 0>$, $Q = <(2, 2, 2); 1, 0>$	[9], $P = Q$	$P > Q$

7.7 MCDM via GSEIFN using the present ranking approach

Here, a methodology of MCDM problem based on the proposed ranking method and proposed IFN has been discussed.

Suppose that $P = \{P_1, P_2, ..., P_n\}$ be a set of m alternatives and $N = \{N_1, N_2, ..., N_m\}$ be a set of n criteria for each alternative P_i. Let $w_c = (w_{c_1}, w_{c_2}, ..., w_{c_m})$ be the weight of criteria where $w_{c_n} \in [0, 1]$ and $w_{c_1} + w_{c_2} + \cdots + w_{c_n} = 1$ to choose the best alternative among $P_i's$ for $i = 1, 2, ..., n$ based on the consummated exponential mean method. The methodology is discussed as follows:

Step 1: Form a table of fuzzy rating criteria by choosing fuzzy criteria in terms of GSEIFNs $p_{ij} = (p_{ij}; w_{ij}, \eta_{ij}), (i = 1, 2, 3, ..., m; j = 1, 2, 3, ..., n)$ and $p_{ij} \in R$ and weight of criteria w_{c_n}.

Step 2: A decision matrix $B = (p_{ij})_{m \times n}$ has been formed by using the above choosing table of fuzzy rating criteria:

$$B = \begin{array}{c} N_1 \\ N_2 \\ \vdots \\ N_n \end{array} \begin{bmatrix} P_1 & P_2 & \cdots & P_n \\ (p_{11}; w_{11}, \eta_{11}) & (p_{12}; w_{12}, \eta_{12}) & \cdots & (p_{1n},; w_{1n}, \eta_{1n}) \\ (p_{21}; w_{21}, \eta_{21}) & (p_{22}; w_{22}, \eta_{22}) & \cdots & (p_{2n}; w_{2n}, \eta_{2n}) \\ \cdots & \cdots & \cdots & \cdots \\ (p_{m1}; w_{m1}, \eta_{m1}) & (b_{m1}; w_{m2}, \eta_{m2}) & \cdots & (b_{mn}; w_{mn}, \eta_{mn}) \end{bmatrix}$$

Step 3: If the given data are inadequate, then normalize the choosing decision matrix $B = (b_{ij})_{m \times n}$ by using the equation

$$a_{ij} = \frac{w_{c_i} p_{ij}}{100} \tag{7.9}$$

Then the normalized decision matrix will be of the form

$$R = \begin{matrix} \\ N_1 \\ N_2 \\ \vdots \\ N_m \end{matrix} \begin{bmatrix} P_1 & P_2 & \cdots & P_n \\ (a_{11}; w'_{11}, \eta'_{11}) & (a_{12}; w'_{12}, \eta'_{12}) & \cdots & (a_{1n}; w'_{1n}, \eta'_{1n}) \\ (a_{21}; w'_{21}, \eta'_{21}) & (a_{22}; w'_{22}, \eta'_{22}) & \cdots & (a_{2n}; w'_{2n}, \eta'_{2n}) \\ \cdots & \cdots & \vdots & \cdots \\ (a_{m1}; w'_{m1}, \eta'_{m1}) & (a_{m1}; w'_{m2}, \eta'_{m2}) & \cdots & (a_{mn}; w'_{mn}, \eta'_{mn}) \end{bmatrix}$$

Step 4: After getting the normalized decision matrix $R = (a_{ij})_{m \times n}$, evaluate $R(P_i)$ by using the following equation:

$$R(P_i) = \sum_{k=1}^{m} M(P_i)_{N_k}, i = 1, 2, \ldots, n \tag{7.10}$$

where $M(P_i)_{N_k}$ is the consummated exponential mean-based rank of GPIFNs.

Step 5: The decision makers choose the alternative of maximum $R(P_i)$.

7.7.1 Numerical example of MCDM

In this section, an MCDM problem has been discussed to exhibit the validity and applicability of the present ranking method and proposed IFN. Suppose that a student searching for a good college among available colleges, namely, P_1, P_2, P_3 for him located in a city on the basis of four criteria represented by GSEIFNs:
(1) $N_1 =$ Transportation
(2) $N_2 =$ Yearly cost
(3) $N_3 =$ Teaching quality
(4) $N_4 =$ Yearly performance

Let the weight of criteria $w_c = (0.1, 0.2, 0.3, 0.4)$ and the fuzzy rating of criteria of each alternative as given in Table 7.2.

The decision matrix $B = (b_{ij})$ is formed by using Table 7.2:

Table 7.2: Rating of criteria.

	P_1	P_2	P_3
N_1	$(40, 20, 30; 0.2, 0.3)$	$(90, 40, 50; 0.4, 0.5)$	$(50, 30, 40; 0.3, 0.4)$
N_2	$(30, 10, 20; 0.1, 0.2)$	$(40, 20, 30; 0.2, 0.3)$	$(60, 40, 50; 0.2, 0.5)$
N_3	$(50, 30, 40; 0.3, 0.4)$	$(80, 30, 40; 0.4, 0.5)$	$(90, 40, 50; 0.4, 0.5)$
N_4	$(60, 40, 50; 0.2, 0.5)$	$(70, 20, 50; 0.3, 0.5)$	$(90, 40, 50; 0.4, 0.5)$

Table 7.3: Table of ranking result by the proposed method.

Alternatives of decision-making problem	Results by using proposed method
P_1	8.5352206
P_2	8.7712329
P_3	8.9213067

$$B = \begin{bmatrix} (40, 20, 30; 0.2, 0.3) & (90, 40, 50; 0.4, 0.5) & (50, 30, 40; 0.3, 0.4) \\ (30, 10, 20; 0.1, 0.2) & (40, 20, 30; 0.2, 0.3) & (60, 40, 50; 0.2, 0.5) \\ (50, 30, 40; 0.3, 0.4) & (80, 30, 40; 0.4, 0.5) & (90, 40, 50; 0.4, 0.5) \\ (60, 40, 50; 0.2, 0.5) & (70, 20, 50; 0.3, 0.5) & (90, 40, 50; 0.4, 0.5) \end{bmatrix}$$

Now normalized the above matrix by using eq. (7.9) as follows:

$$R = \begin{bmatrix} (0.04, 0.02, 0.03; 0.0002, 0.0003) & (0.09, 0.04, 0.05; 0.0004, 0.0005) \\ (0.06, 0.02, 0.04; 0.0002, 0.0004) & (0.08, 0.04, 0.06; 0.0004, 0.0006) \\ (0.15, 0.09, 0.12; 0.0009, 0.0012) & (0.24, 0.09, 0.12; 0.0012, 0.0015) \\ (0.24, 0.16, 0.2; 0.0008, 0.002) & (0.28, 0.08, 0.2; 0.0012, 0.002) \end{bmatrix.$$

$$\begin{matrix} (0.05, 0.03, 0.04; 0.0003, 0.0004) \\ (0.12, 0.08, 0.1; 0.004, 0.001) \\ (0.27, 0.12, 0.15; 0.0012, 0.0015) \\ (0.36, 0.16, 0.2; 0.0016, 0.002) \end{matrix}\right]$$

By using eq. (7.10), result of the decision-making problem is obtained.

By observing Table 7.3 we have $P_3 > P_2 > P_1$; therefore, P_3 college will be chosen by the student.

7.7.2 Conclusion

GSEIFN has been introduced in this chapter. Here we performed some arithmetic operations via (α,β)-cut technique and it is exhibited that fundamental operations on GSEIFNs also give GSEIFN. There are several types of IFNs that have been developed by many researchers. In most of the MCDM problems, researchers used the data or information that are in terms of trapezoidal and triangular IFNs. But in some cases, these IFNs give incorrect solutions by using NTrIFN, NTIFN, and NSEIFN. In this chapter, we have shown that the proposed IFN overcomes these limitations in NTrIFN, NTIFN, and NSEIFN, although a new ranking method has been developed by using the exponential mean of MF and NMF and comparing this method with the other existing methods with the help of some existing ranking approaches and numerical examples. Finally, an MCDM problem has been done by using the proposed IFN and the present novel ranking method to showcase the validity and applicability.

References

[1] Zadeh L.A. (1965) Fuzzy sets, Information and Control, 8(3), 338–353.
[2] Atanassov K.T. Intuitionistic fuzzy sets, In: Intuitionistic fuzzy sets, Physica, Heidelberg, 1999, 1–137.
[3] Prakash K.A., Suresh M., & Vengataasalam S. (2016) A new approach for ranking of intuitionistic fuzzy numbers using a centroid concept, Mathematical Sciences, 10(4), 177–184.
[4] Sudha A.S. & Vijayalakshmi K. (2017) A value and ambiguity-based ranking of symmetric hexagonal intuitionistic fuzzy numbers in decision making, Advanced Fuzzy Mathematics, 12(4), 867–879.
[5] Dutta P. & Saikia B. (2019) Arithmetic operations on normal semielliptic intuitionistic fuzzy numbers and their application in decision-making, Granular Computing, 1–17.
[6] Li D.F., Nan J.X., & Zhang M.J. (2010) A ranking method of triangular intuitionistic fuzzy numbers and application to decision making, International Journal of Computational Intelligence Systems, 3(5), 522–530.
[7] Das S. & Guha D. (2013) Ranking of intuitionistic fuzzy number by centroid point, Journal of Industrial and Intelligent Information, 1, 2.
[8] Roseline S. & Amirtharaj H. (2015) Methods to find the solution for the intuitionistic fuzzy assignment problem with ranking of intuitionistic fuzzy numbers, International Journal of Innovative Research in Science, Engineering and Technology, 4, 7.
[9] Darehmiraki M. (2019) A novel parametric ranking method for intuitionistic fuzzy numbers, Iranian Journal of Fuzzy Systems, 16, 1.
[10] Jianqiang W. & Zhong Z. (2009) Aggregation operators on intuitionistic trapezoidal fuzzy number and its application to multi-criteria decision making problems, Journal of Systems Engineering and Electronics, 20(2), 321–326.
[11] Velu L.G.N., Selvaraj J., & Ponnialagan D. (2017) A new ranking principle for ordering trapezoidal intuitionistic fuzzy numbers, Complexity.

[12] Rezvani S. (2015) Ranking generalized exponential trapezoidal fuzzy numbers based on variance, Applied Mathematics and Computation, 262, 191–198.

[13] Garg H. & Kumar K. (2019) Improved possibility degree method for ranking intuitionistic fuzzy numbers and their application in multiattribute decision-making, Granular Computing, 4(2), 237247.

[14] Nehi H.M. (2010) A new ranking method for intuitionistic fuzzy numbers, International Journal of Fuzzy Systems, 12(1), 8086.

[15] Li D.F. (2010) A ratio ranking method of triangular intuitionistic fuzzy numbers and its application to madm problems, Computers & Mathematics with Applications, 60(6), 1557–1570.

[16] Dubey D. & Mehra A. Linear programming with triangular intuitionistic fuzzy number, 2011, 563–569.

[17] Rezvani S. (2013) Ranking method of trapezoidal intuitionistic fuzzy numbers, Annals of Fuzzy Mathematics and Informatics, 5(3), 515–523.

[18] Zeng X., Li D., & Yu G. 2014 A value and ambiguity-based ranking method of trapezoidal intuitionistic fuzzy numbers and application to decision making, The Scientific World Journal.

[19] Li D.F. & Yang J. (2015) A difference-index based ranking method of trapezoidal intuitionistic fuzzy numbers and application to multiattribute decision making, Mathematical and Computational Applications, 20(1), 25–38.

[20] Das D. & De P.K. (2015) On ranking of trapezoidal intuitionistic fuzzy numbers and its application to multi attribute group decision making, Journal of New Theory(6), 99–108.

[21] Bharati S. (2017) Ranking method of intuitionistic fuzzynumbers, Global Journal of Pure and Applied Mathematics, 13(9), 4595–4608.

[22] Beaula T. & Priyadharsini M. (2015) Fuzzy transportation problem with the value and ambiguity indices of trapezoidal intuitionistic fuzzy numbers, Malaya Journal, Matematik S, (2), 427437.

[23] Jamkhaneh E.B. (2016) A value and ambiguity-based ranking method of generalized intuitionistic fuzzy numbers, Research and Communications in Mathematics and Mathematical Sciences, 6(2), 89103.

[24] Nayagam V.L.G., Jeevaraj S., & Dhanasekaran P. (2018). An improved ranking method for comparing trapezoidal intuitionistic fuzzy numbers and its applications to multicriteria

[25] Xu Z. (2007) Intuitionistic fuzzy aggregation operators, IEEE Transactions on Fuzzy Systems, 15(6), 1179–1187.

[26] De P.K. & Das D. (2012). Ranking of trapezoidal intuitionistic fuzzy numbers. 12th International Conference on Intelligent Systems Design and Applications(ISDA), Kochi, India, pp.184–188

[27] Wu J. & Cao Q. (2013) Same families of geometric aggregation operators with intuitionistic trapezoidal fuzzy numbers, Applied Mathematical Modelling, 37(1–2), 318327.

[28] Nayagam V.L.G., Jeevaraj S., & Sivaraman G. (2016) Complete ranking of intuitionistic fuzzy numbers, Fuzzy Information and Engineering, 8(2), 237–254.

Laxminarayan Sahoo

8 Method for solving intuitionistic fuzzy assignment problem

Abstract: The method to find an answer for assignment problem (AP) under intuitionistic fuzzy domain is proposed in this chapter. Due to the irregular rising and falling of the present market economy, here we have assumed that the assignment costs are not always fixed. Therefore, the assignment costs are imprecise in nature. In the existing literature, different approaches have been used, which are interval, fuzzy, stochastic, and fuzzy–stochastic approaches to represent the impreciseness. In this chapter, we have represented impreciseness taking intuitionistic fuzzy numbers (IFN). The proposed method is hinged on ranking of IFN and use of well-known Hungarian method. Here, we have used a newly proposed centroid concept ranking method for IFNs. In this chapter, we have solved AP where costs for assignment are taken as triangular IFNs. A numerical example has been considered to derive the optimal result and also to adorn the applicability of the suggested method. In the end, concluding remarks and future research of the proposed approach have been presented.

Keywords: assignment problem, intuitionistic fuzzy numbers, Hungarian method, impreciseness, triangular intuitionistic fuzzy numbers

8.1 Introduction

Assignment problem (AP) is effectively used in solving realistic problems for finding the best alternative solutions. AP plays an essential role in managerial decision-making. It is related to the problem of production planning, scheduling, and engineering design problem. The basic task of AP is to allocate many tasks/jobs in a more efficient way such that ideal assignment can be obtained in an acceptable limit. In an AP, n workers and n jobs and effectiveness of each worker for each job, the problem involves in assigning each worker to only one job so that the total amount of effectiveness is optimized. In reality, different procedures and algorithms, namely, linear programming method [1–3], Hungarian algorithm [4], neural network method [5], genetic algorithm [6], and so on were used to solve the AP. In traditional AP, it is assumed that all the parameters related to AP are fixed valued. Notwithstanding in reality, these parameters are not fixed due to uncertainty and lack of proper information. The uncertainty may occur resulting from some factors: (i) decision maker has

Laxminarayan Sahoo, Department of Computer and Information Science, Raiganj, West Bengal, India

https://doi.org/10.1515/9783110628616-008

no sense about the assignment parameters when a job is to be scheduled at the beginning time, as a result some uncertainty may occur with regard to assignment parameters; (ii) nowadays present market economy is always flooded, that is, always unstable due to competing market and, as a consequence, the assignment cost is totally changeable. To succeed in dealing with these types of condition, the problem can be framed using the notion of uncertainty and parameters are served as imprecise/uncertain in nature. In such possible situations, fuzzy set (FS) theory plays an important role in picking up such situation. The theory of FS was originated by Zadeh [7] in 1965 and it dealt with imprecision and vagueness in real-world situations. In the year 1970, Belmann and Zadeh [8] introduced the notion of decision-making problems assuming uncertainty. Here, we have served imprecise parameters considering fuzzy numbers. Therefore, the fuzzy AP (FAP) yields an efficient framework that solves real-life problems with uncertain information. In the last few years, several efforts have been made in the existing relevant works for solving FAP. A number of researchers [9–15] have formulated AP in different sectors/areas such as firm management, vehicle routing, and network management considering fuzzy parameters and/or interval parameters. For solving these problems, they have also used several mathematical techniques/methods like linear programming technique and genetic algorithm. Chen and Chen [16] described about fuzzy optimal solution of the AP based on the ranking of generalized fuzzy numbers. Fuzzy risk analysis based on ranking fuzzy numbers using α-cuts was proposed by Chen and Wang. Rommelfanger [17] discussed a ranking of fuzzy cost present in the FAP, which takes more advantages over the existing fuzzy ranking methods. Kar et al. [18] unfolded the fuzzy generalized AP with restriction on available cost. Solution of FAP with ranking of generalized trapezoidal fuzzy numbers was resolved by Thangavelu et al. [19]. Andal et al. [20] discussed a technique to minimize cost flow of FAP considering fuzzy membership. Pramanik and Biswas [21] introduced multiobjective AP with generalized trapezoidal fuzzy numbers. Mukherjee and Basu [22] suggested the application of fuzzy ranking method for solving AP with fuzzy cost. Very recently, Sahoo and Ghosh [23] solved AP with linguistic costs.

In fuzzy optimization technique, the degrees of receiving objective functions and constraints are considered. FS theory has also been evolved in several fields and its different modifications/upgradations have come to light. The most important generalizations of FS theory is an intuitionistic FS (IFS) which was originated by Atanassov [24] in 1986. The concept of IFS is an alternative approach to define FS in the case where information is not available sufficiently for defining impreciseness by means of ordinary FS. An IFS is characterized by the degree of acceptance and degree of rejection. The IFS theory has been applied in different fields including decision-making. For more details, one may refer to the works of Jana and Roy [25], Roy et al. [26], and Kumar and Hussian [27].

In this chapter, we have considered intuitionistic FAP (IFAP). To solve the IFAP, we have used a new ranking method based on centroid concept introduced by Arun

Prakash et al. [28]. Then the AP has been remodeled into an AP whose parameters are fixed/crisp valued and solved by linear programming method and/or Hungarian method [4]. An example has been studied to obtain the optimal result and also to illustrate the applicability of the proposed method. The rest of the chapter is presented as follows:

In Section 8.2, brief notions about IFSs and fuzzy numbers are described. Mathematical formulation of AP is given in Section 8.3. Section 8.4 gives the solution methodology of the IFAP. In Section 8.5, a numerical example is given for illustration purpose. Section 8.6 concludes the chapter with further scope of research.

8.2 Preliminaries

In this section, a brief idea about IFSs and fuzzy numbers used in this chapter is presented.

Definition 8.1 ([23]). Let X be a given nonempty set. An IFS $\tilde{A}$ in X is defined by $\tilde{A} = \{(x, \mu_{\tilde{A}}(x), \gamma_{\tilde{A}}(x)) : x \in X\}$ where $\mu_{\tilde{A}}(x): X \to [0,1]$ and $\gamma_{\tilde{A}}(x): X \to [0,1]$ define, respectively, the degree of acceptance and degree of rejection of the element $x \in X$ to the set $\tilde{A}$ which is a subset of X and every $x \in X$, $0 \le \mu_{\tilde{A}}(x) + \gamma_{\tilde{A}}(x) \le 1$. Again each IFS $\tilde{A}$ in X, the function $\pi_{\tilde{A}}(x) = 1 - \mu_{\tilde{A}}(x) - \gamma_{\tilde{A}}(x)$ is called the degree of hesitation. If $\pi_{\tilde{A}}(x) = 0$ for all $x \in X$ then IFS becomes FS.

Definition 8.2. An IFS $\tilde{A}$ is called normal if there exists $x_0 \in X$ such that $\mu_{\tilde{A}}(x_0) = 1$ and $\gamma_{\tilde{A}}(x_0) = 0$.

Definition 8.3. An IFS $\tilde{A}$ is called convex if for all $x_1, x_2 \in X$ and $0 \le \lambda \le 1 \mu_{\tilde{A}}(\lambda x_1 + (1-\lambda)x_2) \ge \text{Min}\{\mu_{\tilde{A}}(x_1), \mu_{\tilde{A}}(x_2)\}$ and $\gamma_{\tilde{A}}(\lambda x_1 + (1-\lambda)x_2) \ge \text{Min}\{\gamma_{\tilde{A}}(x_1), \gamma_{\tilde{A}}(x_2)\}$.

Definition 8.4. Support of IFS $\tilde{A}$ with universal set X is denoted by Support $(\tilde{A})$ and is defined by Support $(\tilde{A}) = \{x: \mu_{\tilde{A}}(x) > 0 \text{ and } \gamma_{\tilde{A}}(x) \le 1,\ x \in X\}$.

Definition 8.5. An IFS $\tilde{A} = \{(x, \mu_{\tilde{A}}(x), \gamma_{\tilde{A}}(x)) : x \in X\}$of the real line is called an IFN if
(i) $\tilde{A}$ is intuitionistic fuzzy (IF) normal;
(ii) $\tilde{A}$ is IF convex;
(iii) $\mu_{\tilde{A}}(x)$ is upper semicontinuous and $\gamma_{\tilde{A}}(x)$ is lower semicontinuous; and
(iv) Support$(\tilde{A})$ is bounded.

Definition 8.6. An IFN $\tilde{A} = (a_1, a_2.a_3)(\bar{a}_1, a_2, \bar{a}_3)$ where $\bar{a}_1 \le a_1 \le a_2 \le a_3 \le \bar{a}_3$ is said to be triangular IFN if its membership and nonmembership functions, respectively, are defined as follows:

$$\mu_{\tilde{A}}(x) = \begin{cases} \frac{x-a_1}{a_2-a_1} & \text{if } a_1 \le x \le a_2 \\ \frac{a_3-x}{a_3-a_2} & \text{if } a_2 \le x \le a_3 \\ 0 & \text{otherwise} \end{cases}$$

and

$$\gamma_{\tilde{A}}(x) = \begin{cases} \frac{a_2-x}{a_2-\bar{a}_1} & \text{if } \bar{a}_1 \le x \le a_2 \\ \frac{x-a_2}{\bar{a}_3-a_2} & \text{if } a_2 \le x \le \bar{a}_3 \\ 1 & \text{otherwise} \end{cases}$$

Definition 8.7 ([27]). The centroid point $(\tilde{x}(A), \tilde{y}(A))$ of triangular IFN $\tilde{A} = (a_1, a_2.a_3)$ $(\bar{a}_1, a_2, \bar{a}_3)$ can be determined by

$$x_\mu(A) = \frac{a_1 + a_2 + a_3}{3}$$

$$x_\gamma(A) = \frac{2\bar{a}_1 - a_2 + \bar{a}_3}{3}$$

$$y_\mu(A) = \frac{1}{3}$$

and

$$y_\gamma(A) = \frac{2}{3}$$

Definition 8.8 ([27]). The ranking function of the triangular IFN $\tilde{A}$ is defined by

$$\rho(\tilde{A}) = \left(\frac{1}{2}(x_\mu(A) - y_\mu(A))^2 + \frac{1}{2}(x_\gamma(A) - y_\gamma(A))^2\right)^{\frac{1}{2}}.$$

Theorem 8.1. *Let* $\tilde{A}$ and $\tilde{B}$ *be two triangular IFNs then* $\rho(\tilde{A}+\tilde{C}) > \rho(\tilde{B}+\tilde{C}) \Rightarrow \tilde{A}+\tilde{C} \ge \tilde{B}+\tilde{C}$.

Proof. $\rho(\tilde{A}+\tilde{B}) = \rho(\tilde{A}) + \rho(\tilde{B})$ and $\rho(\tilde{B}+\tilde{C}) = \rho(\tilde{B}) + \rho(\tilde{C})$
Now if, $\tilde{A} \ge \tilde{B}$ then $\rho(\tilde{A}+\tilde{C}) > \rho(\tilde{B}+\tilde{C}) \Rightarrow \tilde{A}+\tilde{C} \ge \tilde{B}+\tilde{C}$.

Theorem 8.2. *Let* $\tilde{A}$ and $\tilde{B}$ *be two triangular IFNs, then,*
$\tilde{A} < \tilde{B}$ *if* $\rho(\tilde{A}) < \rho(\tilde{B})$
$\tilde{A} > \tilde{B}$ *if* $\rho(\tilde{A}) > \rho(\tilde{B})$
$\tilde{A} = \tilde{B}$ *if* $\rho(\tilde{A}) = \rho(\tilde{B})$

8.3 Mathematical formulation of assignment problem

Suppose there are n jobs to be performed and n workers/machines are available for doing these jobs. Here we have assumed that each worker/machine can do each job at a time, though with swing degree of efficiency. Let c_{ij} be the cost or time if the ith person/machine is assigned the jth job, the problem is to find an assignment so that the entire cost of payment or time for performing all jobs are minimum.

Mathematically, AP can be modeled as follows:

$$\text{Minimize } z = \sum_{i=1}^{n}\sum_{j=1}^{n} c_{ij}x_{ij}, \quad (i, j = 1, 2, \ldots, n)$$

$$\text{Subject to } x_{ij} = \begin{cases} 1 & \text{if } i\text{th worker/machine is assigned } j\text{th job} \\ 0 & \text{if } i\text{th worker/machine is not assigned } j\text{th job} \end{cases} \tag{8.1}$$

$$\sum_{j=1}^{n} x_{ij} = 1, \quad i = 1\,(1)\,n$$

and

$$\sum_{i=1}^{n} x_{ij} = 1, \quad j = 1\,(1)\,n$$

where x_{ij} denotes that jth job is to be assigned to the ith worker/machine.

In most of the cases, it is observed that the cost of payment/time of assignment is not precise due to the fluctuation of market economy and, hence, we have assumed that the assignment costs/times are imprecise and have also considered that they are fuzzy numbers $\tilde{c}_{ij}(i = 1\,(1)\,n$ and $j = 1\,(1)\,n)$. Then AP (8.1) becomes FAP, which can be written as follows:

$$\text{Minimize } \tilde{z} = \sum_{i=1}^{n}\sum_{j=1}^{n} \tilde{c}_{ij}x_{ij}, \quad (i, j = 1, 2, \ldots, n)$$

$$\text{Subject to } x_{ij} = \begin{cases} 1 & \text{if } i\text{th worker/machine is assigned } j\text{th job} \\ 0 & \text{if } i\text{th worker/machine is not assigned } j\text{th job} \end{cases} \tag{8.2}$$

$$\sum_{j=1}^{n} x_{ij} = 1, \quad i = 1\,(1)\,n$$

and

$$\sum_{i=1}^{n} x_{ij} = 1, \quad j = 1\,(1)\,n$$

where x_{ij} denotes that jth job is to be assigned to the ith person/machine.

Theorem 8.3. *If* $x_{ij} = X_{ij}$ *minimizes* $\tilde{z} = \sum_{i=1}^{n}\sum_{j=1}^{n} \tilde{c}_{ij}x_{ij}$ *over all* $x_{ij} = 0$ or 1 *such that* $\sum_{j=1}^{n} x_{ij} = 1$, $\sum_{j=1}^{n} x_{ij} = 1$ *then* $x_{ij} = X_{ij}$ *also minimizes* $\tilde{z}' = \sum_{i=1}^{n}\sum_{j=1}^{n} \tilde{c}'_{ij}x_{ij}$ *where* $\tilde{c}'_{ij} = \tilde{c}_{ij} - \tilde{u}_i - \tilde{v}_j$

Proof. Let $x_{ij} = X_{ij}$ minimizes the total cost $\tilde{z} = \sum_{i=1}^{n}\sum_{j=1}^{n} \tilde{c}_{ij}x_{ij}$ over all $x_{ij} = 0$ or 1 and $\sum_{j=1}^{n} x_{ij} = 1$, $\sum_{j=1}^{n} x_{ij} = 1$.

Now it is to be shown that $x_{ij} = X_{ij}$ also minimizes a new cost $\tilde{z}' = \sum_{i=1}^{n}\sum_{j=1}^{n} (\tilde{c}_{ij} - \tilde{u}_i - \tilde{v}_j)$ x_{ij} where $\tilde{u}_i$ and $\tilde{v}_j$ are imprecise cost constants subtracted from cost matrix $(\tilde{c}_{ij})_{n\times n}$.

Again $\tilde{z}' = \sum_{i=1}^{n}\sum_{j=1}^{n} \tilde{c}_{ij}x_{ij}\sum_{j=1}^{n} \tilde{c}_{ij}x_{ij} - \sum_{i=1}^{n}\sum_{j=1}^{n} \tilde{u}_i x_{ij} - \sum_{i=1}^{n}\sum_{j=1}^{n} \tilde{v}_j x_{ij} = z - \sum_{i=1}^{n} \tilde{u}_i \sum_{j=1}^{n} x_{ij} - \sum_{i=1}^{n} x_{ij} \sum_{j=1}^{n} \tilde{v}_j = z - \sum_{i=1}^{n} \tilde{u}_i - \sum_{j=1}^{n} \tilde{v}_j$

So, here $\tilde{z}'$ is independent of x_{ij}'s, and hence $\tilde{z}'$ is minimized when $\tilde{z}$ is minimized and vice versa.

Theorem 8.4. *In an AP with cost matrix* $(\tilde{c}_{ij})_{n\times n}$, *if all* $\tilde{c}_{ij} \geq \tilde{0}$ *then a feasible solution* x_{ij} *is optimal when* $\sum_{i=1}^{n}\sum_{j=1}^{n} \tilde{c}_{ij}x_{ij} = \tilde{0}$.

Proof. Since all $\tilde{c}_{ij} \geq \tilde{0}$ and all $x_{ij} \geq 0$, therefore $\left(\tilde{z} = \sum_{i=1}^{n}\sum_{j=1}^{n} \tilde{c}_{ij}x_{ij}\right) \geq \tilde{0}$. Hence, the minimum possible value of $\tilde{z} = \sum_{i=1}^{n}\sum_{j=1}^{n} \tilde{c}_{ij}x_{ij}$ can attain is always $\tilde{0}$. Then any feasible solution x_{ij} is optimal iff $\sum_{i=1}^{n}\sum_{j=1}^{n} \tilde{c}_{ij}x_{ij} = \tilde{0}$.

8.3.3 The Hungarian method

The following method applies Theorem 8.4 to a given $n \times n$cost matrix to obtain an optimal assignment [4].

Step 1: For each row $i, i = 1, 2, ..., n$, let $c_{ir}, r \in \{1, 2, ..., n\}$ be the smallest entry and calculate $c_{ij} - c_{ir}$, for each i and $r \in \{1, 2, ..., n\}$.
Obviously $\text{Min}\{c_{ij} - c_{ir}\} = 0$, for each i and $r \in \{1, 2, ..., n\}$.

Step 2: For each column $j, j = 1, 2, ..., n$, let $c_{kj}, k \in \{1, 2, ..., n\}$ be the smallest entry and calculate $c_{ij} - c_{kj}$, for each j and $k \in \{1, 2, ..., n\}$.
Obviously $\text{Min}\{c_{ij} - c_{kj}\} = 0$, for each j and $k \in \{1, 2, ..., n\}$.

Step 3: Find minimum number of lines l (say) in such a way that l l ines covered all the zero entries of the cost matrix.

Step 4: Now, either $l = n$ or $l < n$. If $l = n$, then an optimal assignment is arrived. (ii) If $l < n$ then go to Step 5.

Step 5: Find the smallest entry not covered by any line, subtract this from each uncovered row and column, and then add it to junction entry and go to **Step 3**.

8.4 Solution methodology

Suggested approach of solution procedure of linguistic AP is as given below:

Step 1: Find $\rho(\tilde{c}_{ij})$ of each $\tilde{c}_{ij}$ using ranking function defined in Definition 8.8.

Step 2: Check whether the given AP is balanced or not.
(i) If the AP is a balanced then go to **Step 4.**
(ii) If the AP is unbalanced then go to **Step 3.**

Step 3: Make dummy rows and/or columns with zero cost so as to form a balanced AP.

Step 4: Solved by method A: (existing linear programming problem method) and/or by method B: (the Hungarian method described in Section 8.3.3) to obtain the appropriate assignments.

Method A: Solve

$$\text{Minimize } z = \sum_{i=1}^{n}\sum_{j=1}^{n}\rho(\tilde{c}_{ij})x_{ij}$$

$$\text{subject to } \sum_{i=1}^{n} x_{ij} = 1, \ j = 1, 2, \ldots, n$$

$$\sum_{j=1}^{n} x_{ij} = 1, \ i = 1, 2, \ldots, n$$

$$x_{ij} \in \{0, 1\}$$

Method B: Find an optimal assignment using Hungarian algorithm.

8.5 Numerical example

In this section, we have considered an IFAP with three resources W_1, W_2, W_3 and three jobs J_1, J_2, J_3 with IF assignment cost. The cost matrix $(\tilde{c}_{ij})_{3\times3}$ is given below whose elements are triangular IFNs.

		Activities		
		J_1	J_2	J_3
Resource	W_1	(4,7,14)(3,7,24)	(12,14,16)(11,14,28)	(8,15,25)(7,15,30)
	W_2	(6,18,25)(5,18,28)	(13,19,28)(12,19,36)	(20,25,33)(19,25,38)
	W_3	(20,25,38)(19,25,42)	(2,11,14)(1,11,19)	(6,12,15)(5,12,20)

Solution: Here, it is seen that the IFAP is a balanced one. Now, we have obtained $R(\tilde{c}_{ij})$ of each $\tilde{c}_{ij}$using ranking function mentioned in Definition 8.8. Now the reduced problem is as follows:

		Activities		
		J_1	J_2	J_3
Resource	W_1	12.02	15.57	17.41
	W_2	15.67	22.49	27.38
	W_3	29.58	8.33	11.35

Then, the AP can be modeled in the following linear programming problem as follows:

$$\text{Minimize } z = 12.02x_{11} + 15.67x_{21} + 29.58x_{31} + 15.57x_{12} + 22.49x_{22} + 8.33x_{32} + 17.41x_{13} + 27.38x_{23} + 11.35x_{33} \tag{8.3}$$

Subject to

$$x_{11} + x_{21} + x_{31} = 1;\ \ x_{12} + x_{22} + x_{32} = 1;\ \ x_{13} + x_{23} + x_{33} = 1;$$

$$x_{11} + x_{12} + x_{13} = 1;\ \ x_{21} + x_{22} + x_{23} = 1;\ \ x_{31} + x_{32} + x_{33} = 1;$$

$$x_{ij} \in \{0, 1\},\ i = 1, 2, 3, 4 \text{ and } j = 1, 2, 3$$

Solving problem (8.3) we get the optimal assignments $W_1 \to J_3$, $W_2 \to J_1, W_3 \to J_2$. Also proceeding by the Hungarian method, the optimal assignments are $W_1 \to J_2$, $W_2 \to J_1$, $W_3 \to J_3$. Now, for optimal assignments $W_1 \to J_3$, $W_2 \to J_1, W_3 \to J_2$ the minimum assignment cost in terms of IF is (8,15,25)(7,15,30) + (6,18,25)(5,18,28) + (2,11,14)(1,11,19), that is, (16,44,64)(13, 44, 77) and for optimal assignments $W_1 \to J_2$, $W_2 \to J_1$, $W_3 \to J_3$ the minimum assignment cost in terms of IF is (12,14,16)(11,14,28) + (6,18,25)(5,18,28) + (6,12,15)(5,12,20), that is, (24, 44, 56)(21, 44, 76). More effectively, in reality the assignment cost cannot be taken as fixed/precise one. So, in this chapter we have solved an AP considering each cost of the assignment is IF. For this purpose, to solve the AP with IF cost we have used the ranking function method defined in Definition 8.8 For the IF optimal costs (16, 44, 64)(13, 44, 77), we have obtained $\rho((16, 44, 64)(13, 44, 77)) = 31.95$ and $\rho((24, 44, 56)(21, 44, 76)) = 33.59$. Using, Theorem 8.2, we have seen that $\rho((16, 44, 64)(13, 44, 77)) < \rho((24, 44, 56)(21, 44, 76))$; hence, (16,44,64)(13, 44, 77) is the minimum assignment cost.

8.6 Conclusion

In this chapter, AP with IF assignment cost/time is considered and its solution procedure has been established. AP is one of the most essential problems in decision-making. In everyday situations, parameters of the AP are imprecise. The AP with IF cost is more reasonable than the AP with precise parameters because most of the real-life events are from uncertain domain. The proposed procedure presented here is very effortless and easy for implementation. This solution procedure is based on IF representation of assignment parameters and ranking measure of IF parameters. Here, a newly developed ranking measure based on centroid concept has been used. Then, the AP has been transformed to an AP whose parameters are crisp/precise valued and solved by the existing Hungarian method/linear programming method. Lastly, an IFAP has been solved, and evaluated results have been conferred and measured. It may be asserted that the entire solution procedure set out in this chapter can be put in to solve other realistic decision-making problems in the near future demanding IF parameters.

References

[1] Balinski M.L. (1986) A competitive (dual) simplex method for the assignment problem, Math Program, 34(2), 125–141.

[2] Barr R.S., Glover F., & Klingman D. (1977) The alternating basis algorithm for assignment problems, Math Program, 13(1), 1–13.

[3] Huang M.S. & Rom W.O. (1980) Solving the assignment problem by relaxation, Operations Research, 28(4), 969–982.

[4] Kuhn H.W. (1955) The Hungarian method for assignment problem, Naval Research Logistic Quarterly, 2, 83–97.

[5] Eberhardt S.P., Daud T., Kerns A., Brown T.X., & Thakoor A.P. (1991) Competitive neural architecture for hardware solution to the assignment problem, Neural Networks, 4(4), 431–442.

[6] Avis D. & Devroye L. (1985) An analysis of decomposition heuristic for assignment problem, Operations Research Letter, 3(6), 279–283.

[7] Zadeh L.A. (1965) Fuzzy sets, Information and Control, 8(3), 338–352.

[8] Bellman R.E. & Zadeh L.A. (1970) Decision making in a fuzzy environment, Management Science, 17, 141–164.

[9] Sakawa M., Nishizaki I., & Uemura Y. (2001) Interactive fuzzy programming for two-level linear and linear fractional production and assignment problems: a case study, European Journal of Operational Research, 135, 142–157.

[10] Chen M.S. (1985) On a fuzzy assignment problem, Tamkang Journal, 22, 407–411.

[11] Huang L.S. & Zhang L.-P., Solution method for fuzzy assignment problem with restriction on Qualification. Proceeding of Sixth International Conference on Intelligent Systems Design and Applications (ISDA06), 2006.

[12] Liang-hsuan C. & Hai-wen L. (2007) An extended assignment problem considering multiple inputs and outputs, Applied Mathematical Modeling, 31, 2239–2248.

[13] Majumdar J. & Bhunia A.K. (2007) Elitist genetic algorithm for assignment problem with imprecise goal, European Journal of Operational Research, 177, 684–692.

[14] Ye X. & Xu J. (2008) A fuzzy vehicle routing assignment model with connection network based on priority-based genetic algorithm, World Journal of Modeling and Simulation, 4, 257–268.

[15] Thorani Y.L.P. & Ravi Shankar N. (2013) Fuzzy assignment problem with generalized fuzzy numbers, Applied Mathematical Sciences, 71(7), 3511–3537.

[16] Chen M.S. & Chen J.H. (2009) Fuzzy risk analysis based on the ranking generalized fuzzy numbers with different heights and different spreads, Expert System Applications, 36, 6833–6842.
[17] Rommelfanger H.J. (2004) The advantages of fuzzy optimization models in practical use, Fuzzy Optimization and Decision Making, 3(4), 295–309.
[18] Kar S., Basu K., & Mukherjee S. (2014) Solution of generalized fuzzy assignment problem with restriction on costs under fuzzy environment, International Journal of Fuzzy Mathematics and Systems, 4(2), 169–180.
[19] Thangavelu K., Uthra G., & Umamageswari R.M. (2016) Solution of fuzzy assignment problem with ranking of generalized trapezoidal fuzzy numbers, International Journal of Pure and Applied Mathematics, 106(6), 9–16.
[20] Andal S., Murgesan S., & Ramesh B.K. (2016) A novel approach to minimum cost flow of fuzzy assignment problem with fuzzy membership functions, Engineering, Science and Technology: an International journal, 6(2), 85–88.
[21] Pramanik S. & Biswas P. (2012) Multi-objective assignment problem with generalized trapezoidal fuzzy numbers, International Journal of Applied Information Systems, 2(6), 13–20.
[22] Mukherjee S. & Basu K. (2010) Application of fuzzy ranking method for solving assignment problem with fuzzy costs, International Journal of Computational and Applied Mathematics, 5359–5368.
[23] Sahoo L. & Ghosh S. (2017) Solving assignment problem with linguistic costs, Journal of New Theory, 17, 26–37.
[24] Atanassov K. (1986) Intutionistic fuzzy sets, Fuzzy Sets and Systems, 20, 87–96.
[25] Jana B. & Roy T.K. (2007) Multi-objective intuitionistic fuzzy linear programming and its application in transportation model, Notes Intuitionistic Fuzzy Sets, 13(1), 34–51.
[26] Roy S.K., Ali E., Verdegay J.L., & Das S. (2018) New approach for solving intuitionistic fuzzy multi-objective transportation problem, Sadhana, 43(3), 1–12.
[27] Kumar P.S. & Hussain R.J. (2016) Computationally simple approach for solving fully intuitionistic fuzzy real life transportation problems, International Journal of System Assurance Engineering and Management, 7(1), 90–101.
[28] Arun Prakash K., Suresh M., & Vengataasalam S. (2016) A new approach for ranking of intuitionistic fuzzy numbers using a centroid concept, Mathematical Sciences, 10(4), 177–184.

Goutam Kumar Bose, Pritam Pain

9 Optimization of EDM process through evolutionary computing and fuzzy MCDM techniques

Abstract: This chapter intends to explore the best possible set of process parameters, namely, current, voltage, and pulse on time during electric discharge machining (EDM) process in order to examine the operational changes pertaining to the rate of material removed and the rate with which electrode wears. Here the workpiece selected is high carbon high chromium die steel, and the titanium nitride–coated copper electrode is used. Initially, the data set is trained and validated and finally tested by employing artificial neural network. Then multiobjective genetic algorithm (M-GA) is applied in order to find better parametric combinations for the response values. Finally, multicriteria decision making has been used in order to obtain the noblest parametric combination from the data of M-GA, where two contradictory responses like high material removal rate and low electrode wear rate can be achieved.

Keywords: HCHCr tool, MRR, EWR, ANN, GA, fuzzy, MCDM

9.1 Introduction

As there is advancement in the usage of day-to-day materials, the secondary manufacturing processes have become very essential in order to sustain in that challenging manufacturing industries. Electrical discharge machining (EDM) is extensively popular in the nonconventional machining process, while manufacturing various industrial critical components related to plastic molding, automobile, aerospace, and so on. The process utilizes thermal energy in the form of controlled sparks to remove materials from the conductive workpiece. Due to this distinctive feature of the process, it provides an edge in machining complex contours on high-temperature strength-resistant materials over their contemporary traditional counterparts [1]. In the EDM process, series of sparks are generated from the conductive electrode due to the necessary potential difference created between the narrow tool work gaps. The complete machining zone is flooded with die electric fluid, which simultaneously cools the system as well as helps in flushing the debris produced during machining. The temperature generated during this spark emission across

Goutam Kumar Bose, Pritam Pain, Department of Mechanical Engineering, Haldia Institute of Technology, Haldia, West Bengal, India

https://doi.org/10.1515/9783110628616-009

the tool work gap ranges between 8,000 and 12,000 °C, which is capable enough in melting all types of conductive electrodes that are difficult to machine materials. The parameters that influence the machining process most are duty cycle, pulse off time, pulse on time, current, voltage, spark gap type of electrode, and so on.

Few past research works on EDM have been elaborated here in order to select the proper design of experiment and control parameter. Kanthababu [2] has introduced the basic ideas of single and multiobjective optimization, and also the evolutionary algorithms, as well as gives a brief outline of their relevant characteristics. Some of the metaheuristic algorithm normally applied while solving critical multiobjective functions are shown explicitly. Palanikumar et al. [3] have discussed the optimization of performance characteristics by using the Taguchi method along with an evolutionary gray–fuzzy system with respect to several responses like material removal rate (MRR), surface roughness (Ra), and specific cutting pressure (Pr). Their results show that the optimization technique is very significant while performing machining operations of Glass Fiber Reinforced Polymer (GFRP) composites to achieve good surface quality with minimum low tool wear concurrently. Pandian Vasant [4] during solving of fuzzy programming problems in industrial production systems has operated with three metaheuristic optimization techniques, where a decision maker takes the final call. Here in order to unravel the comprehensive challenges of genetic algorithms (GA) practically, mesh adaptive direct search methods are taken into consideration along with the pattern search. Meyer et al. [5] have investigated problems through distinctive market research with respect to small data sets through statistical techniques. With these small data sets, they have added three different neural network models to this investigation. Deo et al. [6] have predicted the wind speed of 15 different places in Queensland, Australia, by employing artificial neural network (ANN) along with GA. Sahu and Mahapatra [7] have undergone EDM on titanium alloy workpiece. Three different tool materials, namely AlSiMg, copper, and graphite, are used while altering different control parameters. Finally, the surface finish, namely, Ra, Rt, and Rz is determined. Bose et al. [8] have performed optimization of EDM process using artificial bee colony algorithm for multiresponse optimization and whale optimization algorithm in case of single response optimization. Bose and Pain [9] have focused on the optimization of electrochemical grinding process, while machining Alumina–aluminium (Al_2O_3–Al), an interpenetrating phase composite (IPC) by using bioinspired metaheuristics. Prabhu and Kumar [10] have investigated the electrochemical deburring method using nickel-coated tool on Al_6O_{82} material successfully for deburring complex geometrical cross-holes. The input parameters such as current, electrolyte concentration, current density, and time varied in order to evaluate the performance of the tool as well as the analysis of machining characteristics are done.

From the brief literature review, the primary goal of this chapter is to explore the best possible set of process parameters, namely, current (I), voltage (V), and pulse on time (ton) in order to determine the changes in MRR and electrode wear rate (EWR) during EDM of high carbon high chromium die steel (HCHCr) workpiece by applying the titanium nitride–coated copper electrode. Applying the regression analysis, the initial investigation is commenced. Then the data set is trained and validated and finally tested applying ANN. Then multiobjective GA (M-GA) is applied to those data in order to find a better input–output parametric combination. Finally, to find out the noble parametric combination to satisfy the two contradictory responses like high MRR and low EWR, multicriteria decision making (MCDM) is used.

9.2 Experimental design

The experimental investigation is done using die sinking type EDM (ACTSPARK SP1, China make). A positive polarity is maintained while using titanium nitride–coated copper tool electrode. The dielectric used is EDM oil 30 having 0.80 specific gravity (at temp. 25 °C) and 3.11 cSt. viscosity (at 38 °C). The flushing pressure of 0.2 kgf/cm^2 is maintained in between the copper tool ($12 \times 12\ mm^2$) and the workpiece during machining. The tool work gap is made constant by using a servo control system.

The experimental setup is shown in Figure 9.1.

Figure 9.1: Machining setup of EDM process.

Keeping in view of the earlier research works done in this setup of previous experimentation, three control parameters are chosen. These parameters having three levels of variation are considered for 17 experimental runs. The input parameters with their levels of variation are illustrated in Table 9.1. Some of the insignificant factors retained are fixed.

Table 9.1: Process controller with different levels of variation.

Process variables	Notation	Units	Levels of variation		
			L_1	L_2	L_3
Current	I	amp	6	8	10
Voltage	V	volt	45	50	55
Pulse on time	ton	μs	50	100	200

The MRR and EWR are calculated as the weight of the workpiece material removed over time and are expressed as g/min, respectively.

9.3 Regression analysis

Generally, in case of one dependable variable, let us say "*y*" which is dependent on an independent variable "*k*." The correlation among these two types of variables is characterized as a regression model. A strong correspondence exists relating to the design of experiments and regression analysis. A regression model is developed by using a subset of the existing regression which involves contradictory objectives as follows:

- The regression model should incorporate maximum regressors for composing the resultant equation beneficial for the analysis.
- There should be minimum regressors in the regression model in order to lower the cost involved in gathering and sustaining the information comprehensively.

$$\begin{aligned} MRR = & -(-0.016 + 0.0264{*}I - 0.00021{*}V + 0.000074{*}ton - 0.000164{*}I{*}V \\ & + 0.000007{*}I{*}ton - 0.000004{*}V{*}ton) \end{aligned} \tag{9.1}$$

$$\begin{aligned} EWR = & -0.0564 + 0.00675{*}I + 0.000867{*}V + 0.000041{*}ton - 0.000075{*}I{*}V \\ & - 0.000005{*}I{*}ton + 0.000001{*}V{*}ton \end{aligned} \tag{9.2}$$

9.4 Artificial neural network

An ANN is a bioinspired metaheuristic. A neural network consists of nodes that are interconnected thoroughly. Each neuron sends an input signal to the adjacent neuron. It then subsequently exudes the output signals (O_i, $i = 1$ to j) to the subsequent neuron. Every input P_i multiplies by the corresponding weight W_i in synapses and then transfer to the hidden layer as follows:

$$I_i = \sum_{i=1}^{n} W_i P_i \tag{9.3}$$

Output from every node will generate only when the input reaches the threshold level of that node. This output is a sigmoidal function in general as follows

$$O_i = f(I_i) = 1/\left(1 + e^{-I_i}\right) \tag{9.4}$$

Figure 9.2 shows the architecture of a node of ANN.

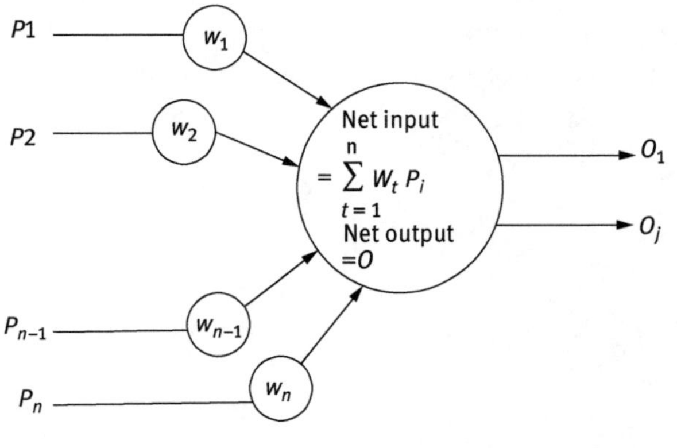

Figure 9.2: Illustration of a synapse of artificial neural network.

The commonly used neural network architectures are as follows:
- Self-organizing networks
- Feedforward networks
- Backpropagation networks

The diagrammatic representation of an ANN has been illustrated in Figure 9.3.

9.4.1 Result analysis applying ANN

The design of experiment is conducted by using soft computing technique through Matlab R2015a. Several algorithms are considered to facilitate the calculation such

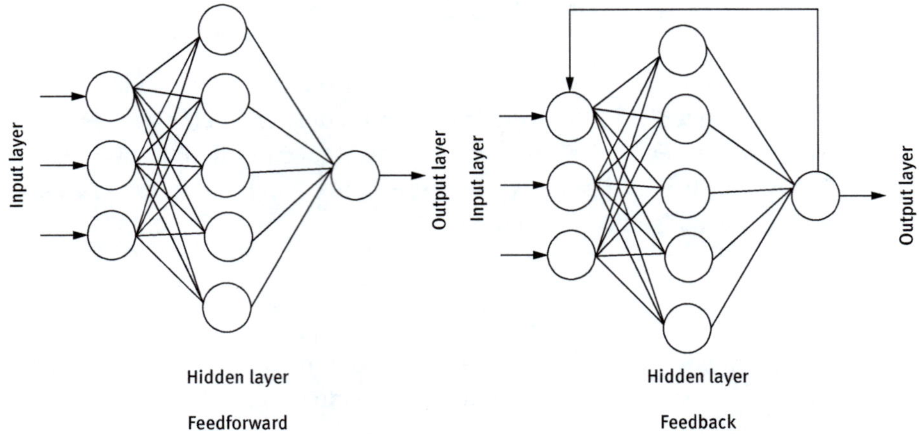

Figure 9.3: Schematic design of a neural network.

as divider and function for dividing the data set, for training Levenberg–Marquardt, and mean squared error (MSE) for performance study. In Figure 9.4, a simulink model is illustrated having the four parameters under consideration in order to optimize the different responses.

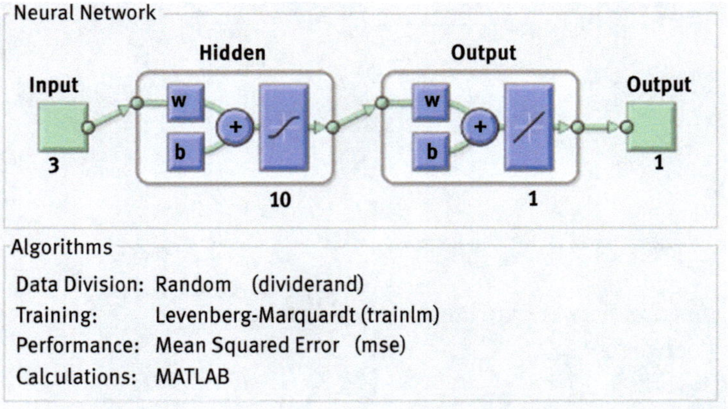

Figure 9.4: Simulink diagram.

The gradient as well as the Jacobian is determined by using backpropagation algorithm techniques.

By applying the chain rule of calculus in the reverse order, the backpropagation computation is achieved in the overall network. Here in the ANN model MSE is formulated between the output of the network and the corresponding target value.

An epoch represents the set of initial value and/or target value (training vectors) in an ANN network which is multiplied by its corresponding weights. Initially 1,000 epochs, which is the maximum number possible, are set to stop the training for all. The graphical representation of the regression displays the correlation among the targets and network outputs. The regression plots illustrate the training, validation, testing, and subsequent overall data. The target, which is the difference between the perfect result and outputs, is illustrated by dashed lines. Whereas the best fit line is illustrated as a continuous line, which signifies the linear regression line relating to the targets and outputs. When a direct correlation exists between the targets and the outputs ten R becomes one, else it becomes zero.

9.4.1.1 MRR analysis

Now while computing using ANN the following data distribution is made. For training, validation, and testing 60%, 15%, and 25% data are taken into consideration, respectively. Here MSE is achieved after three number of iterations. The training will terminate when the value of the gradient is less than 1.00×10^{-7} and when the validation check number is 6. During computation, the gradient observed is 1.5×10^{-9} and the validation checks reach 4 after successive iterations.

Figure 9.5 illustrates the execution of neural network training for MRR.

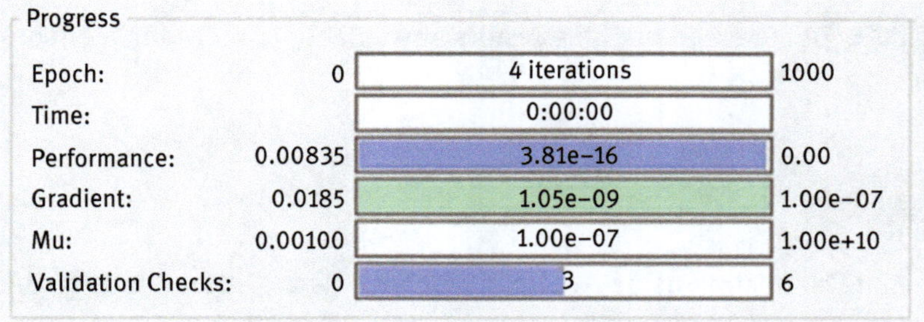

Figure 9.5: Neural network training for MRR.

It is observed that the best validation result is 0.00068473 at epoch 1. The validation performance graph for MRR is shown in Figure 9.6.

It is shown in Figure 9.6 that at 1, the validation experiment is minimum while training continued till 4 before terminating. The plot shows a significant relationship between testing and validation. The test curves and the validation curves indicate close resemblance among themselves.

Figure 9.7 demonstrates the following results of R: training = 0.96358, validation = 1, testing = 0.98834, and overall = 0.86907. Hence, it can be inferred that the

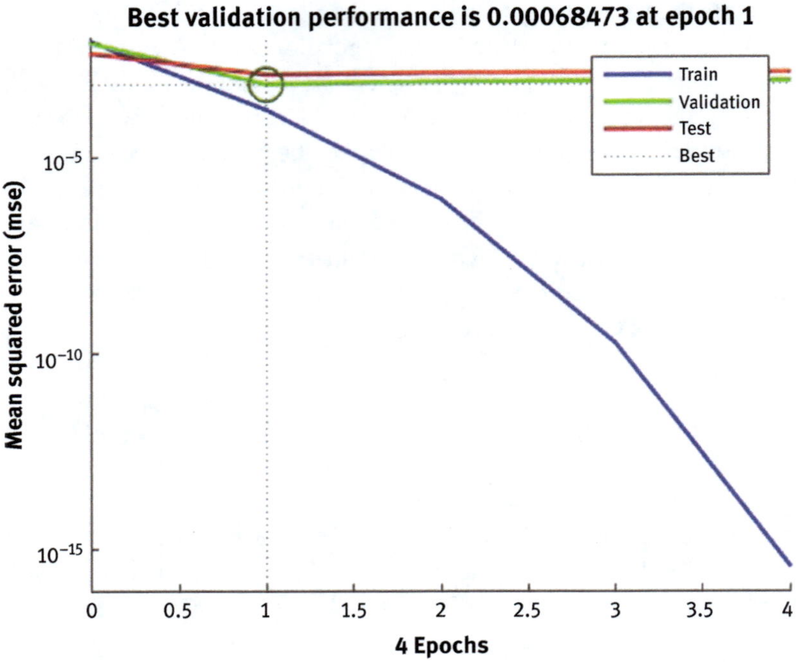

Figure 9.6: Validation performance graph for output MRR.

training data fits good here and both the results of validation and testing are more than 0.9.

9.4.1.2 EWR analysis

At this point while computing using ANN, the following data distribution is made. For training, validation, and testing the data under consideration are 60%, 20%, and 20%, respectively. Here the MSE is achieved after the first iteration. The training will terminate when the value of the gradient is less than 1.00×10^{-7} and when the validation check number reaches 6. During computation gradient observed is 2.68×10^{-8} and the validation checks reaches 1 after successive iterations.

Training for EWR has been illustrated in Figure 9.8.

It is observed that at the fourth epoch, the finest validation result is attained, that is, 1.1425×10^{-5}. Figure 9.9 shows the validation performance for output EWR.

Figure 9.9 shows that at 4 the validation experiment is minimum while training continued till 5 before terminating. The plot highlights a considerable correlation between testing and validation. The test curves and the validation curves show close similarity between them.

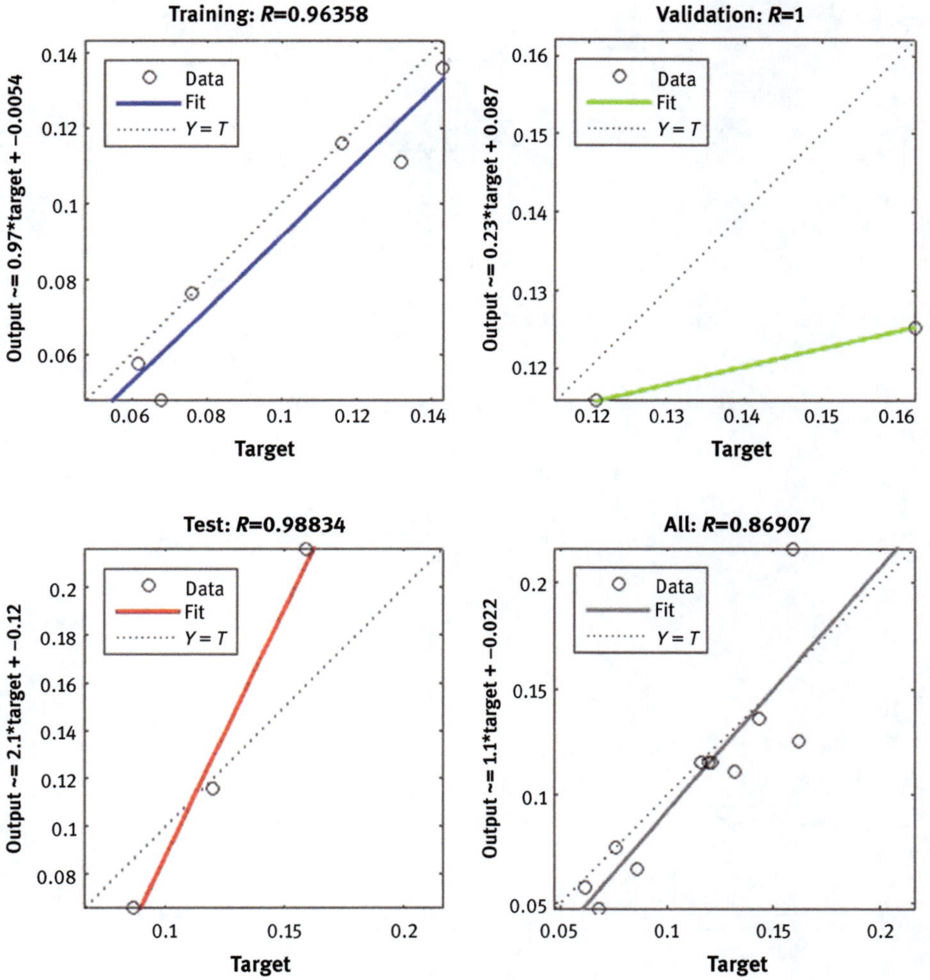

Figure 9.7: Regression graph of output MRR.

Figure 9.10 exhibits the following results of R: training = 0.99986, validation = 1, testing = 1, and overall = 0.97029. Hence, it can be inferred that the training data fits good here and both the results of validation and testing are more than 0.9.

9.5 Genetic algorithm

Depending on the Darwin's theory of survival of the fittest, a metaheuristics approach like GA has been established [11]. It is a nature-inspired metaheuristic that helps in optimizing multifaceted design problems, which incorporates a combination of

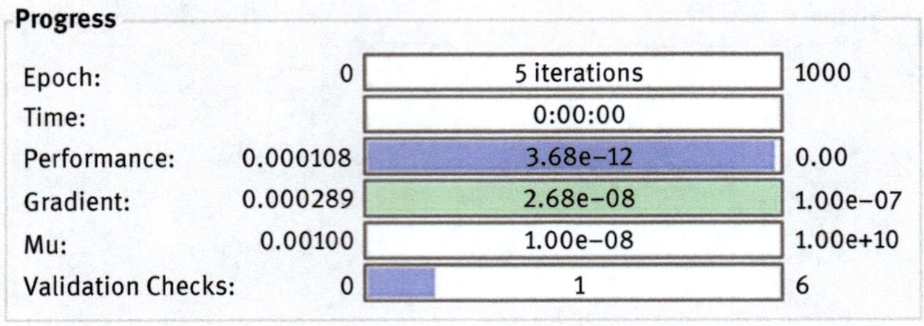

Figure 9.8: Neural network training for the output EWR.

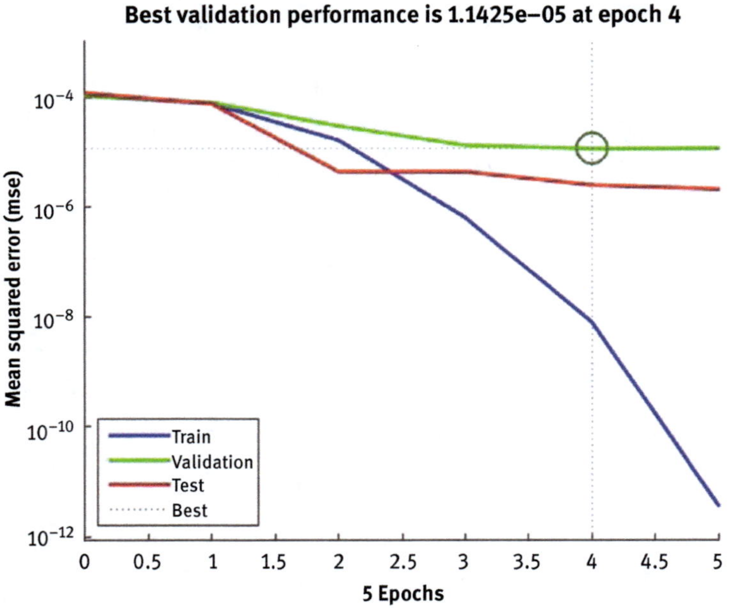

Figure 9.9: Performance graph for the output EWR.

constant discrete variables, irregular and lying in nonconvex design space. It is well competent in locating the optimum solution globally with maximum probability.

The first changes on the original bounded function are done by means of penalty function and transformed it into unbounded function as follows:

$$P = \text{variable} \tag{9.5}$$

$$\text{Minimize } f(P) + R\sum_{i=1}^{n}\Phi(g_i(P)) \tag{9.6}$$

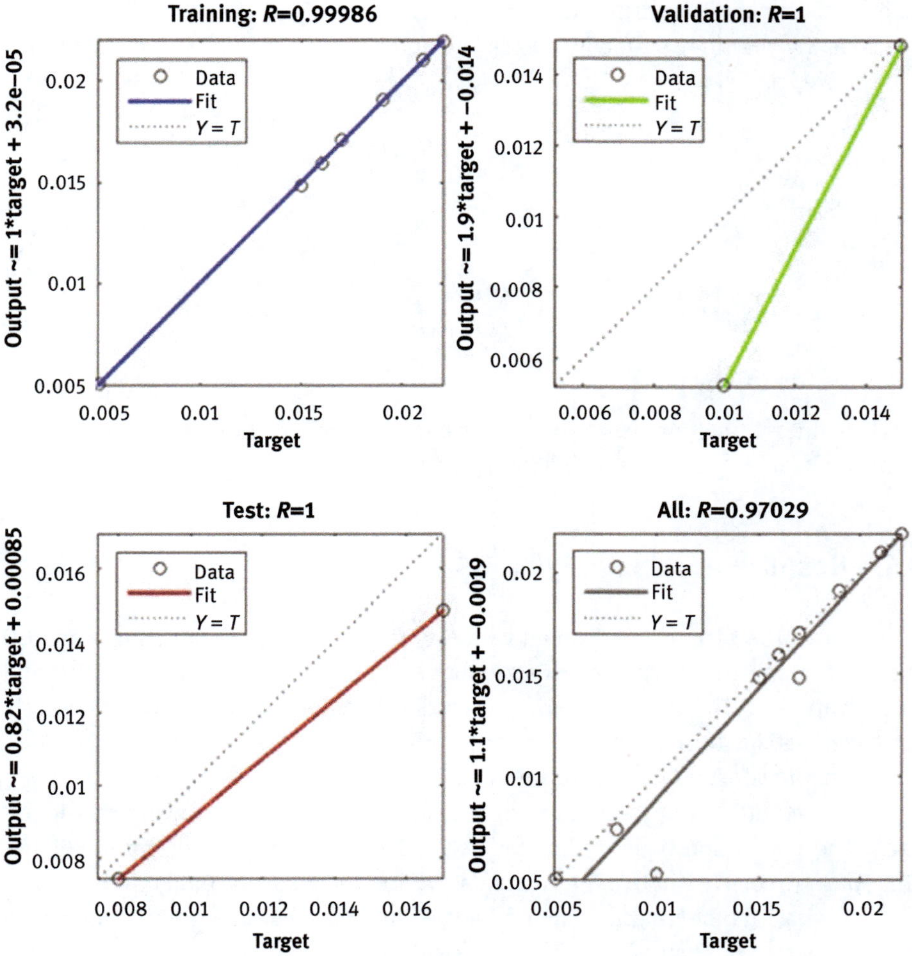

Figure 9.10: Regression plot of EWR.

$$\text{Subjected to } p_x^{(l)} \le p_x \le p_x^{(u)},\ x = 1, 2, \ldots, n \tag{9.7}$$

where Φ denotes the penalty function as follows:

$$\Phi = (X) = \langle X \rangle^2 \tag{9.8}$$

where

$$\langle X \rangle = \begin{Bmatrix} X & \text{if} \quad X > 0 \\ 0 & \text{if} \quad X \le 0 \end{Bmatrix} \tag{9.9}$$

R denotes the penalty constant for the parameters.

Boosting the fitness function, $F(X)$, the subsequent transformation function is minimalized as follows:

$$F(P) = F_{\max} - \left(f(P) + R \sum_{i=1}^{n} \Phi(g_i(P)) \right) = F_{\max} - f'(P) \tag{9.10}$$

where

$$F_{\max} > f'(P) \tag{9.11}$$

$f(P)$ is the fitness function.

This fitness function usually performs either YES or NO function. This arrangement can be looked upon as a genetic chromosome.

9.5.1 Result analysis applying GA

Here dual optimization is performed with the help of the regression equations (9.1) and (9.2), respectively. The response functions, that is, maximization of MRR and minimization of EWR are ambiguous in nature. But they are to be simultaneously optimized satisfactorily. Now the input parameters with their boundary condition are taken into consideration from Table 9.1.

The computation of the algorithm is performed with the following considerations. The population type is a double vector having a population size of 50. The functions are selected in two tournaments. During the analysis, 0.8 is the crossover fraction with 1 as the intermediate crossover function. The forward migration with 0.2 fraction and interval of 20 has been considered. Here, 0.35 fraction is set for the Pareto front population. The stopping criterion by considering 100 stall generations is 100 × numbers of variables, where the stall limit is infinity. The tolerance of the function is 0.0001 with tolerance for the constraint is 0.001.

Therefore, the total number of iterations completed is 158. There are 17 number of parametric combinations along with their objective functions. The final result is shown in Table 9.2. When the average change in the dispersion of the Pareto solution is achieved toward its limiting condition, the program is terminated. The contradictory outputs are optimized simultaneously by this GA, which is represented by various graphs in Figure 9.11. From this figure it is imperative that the control parameters fluctuate within a range. Here MRR value ranges between 0.085 and 0.163 g/min, respectively, while EWR varies between 0.006 and 0.018 g/min, respectively. Therefore, for attaining at an optimal parametric combination where this dual objective gets satisfied, fuzzy gray relational analysis (FGRA) is done.

Table 9.2: Variation of parameters with their responses.

S. no.	Current	Voltage	Pulse on time	MRR	EWR
1	6	45	50	0.085	0.006
2	10	45	52	0.163	0.018
3	9	45	51	0.136	0.014
4	8	45	50	0.116	0.011
5	8	45	51	0.126	0.012
6	7	45	50	0.096	0.007
7	9	45	50	0.152	0.017
8	6	45	51	0.092	0.007
9	9	45	51	0.138	0.014
10	9	45	51	0.151	0.016
11	8	45	52	0.131	0.013
12	8	45	50	0.119	0.011
13	10	45	52	0.158	0.017
14	7	45	51	0.099	0.008
15	10	45	51	0.159	0.018
16	9	45	52	0.142	0.015
17	6	45	51	0.087	0.006

9.6 Multiobjective gray relation analysis

Gray relation analysis (GRA) is a multiobjective optimization technique developed by Deng in 1989. Ever since its inception, it has found its presence in experimental design where the model can be unspecified or the information can be insufficient, thereby reflecting a mixture of familiar and unfamiliar. Thus, GRA bridges the correlation among the expected result (finest/perfect) with the actual experimental findings. Gray grade is the average of the gray coefficient. The calculated gray grade denotes that the relation is robust if the value is toward 1. If the value is toward 0, the stability of the model is very poor.

In order to obtain the maximized performance value for some parameter like MRR, then the equation is normalized as follows:

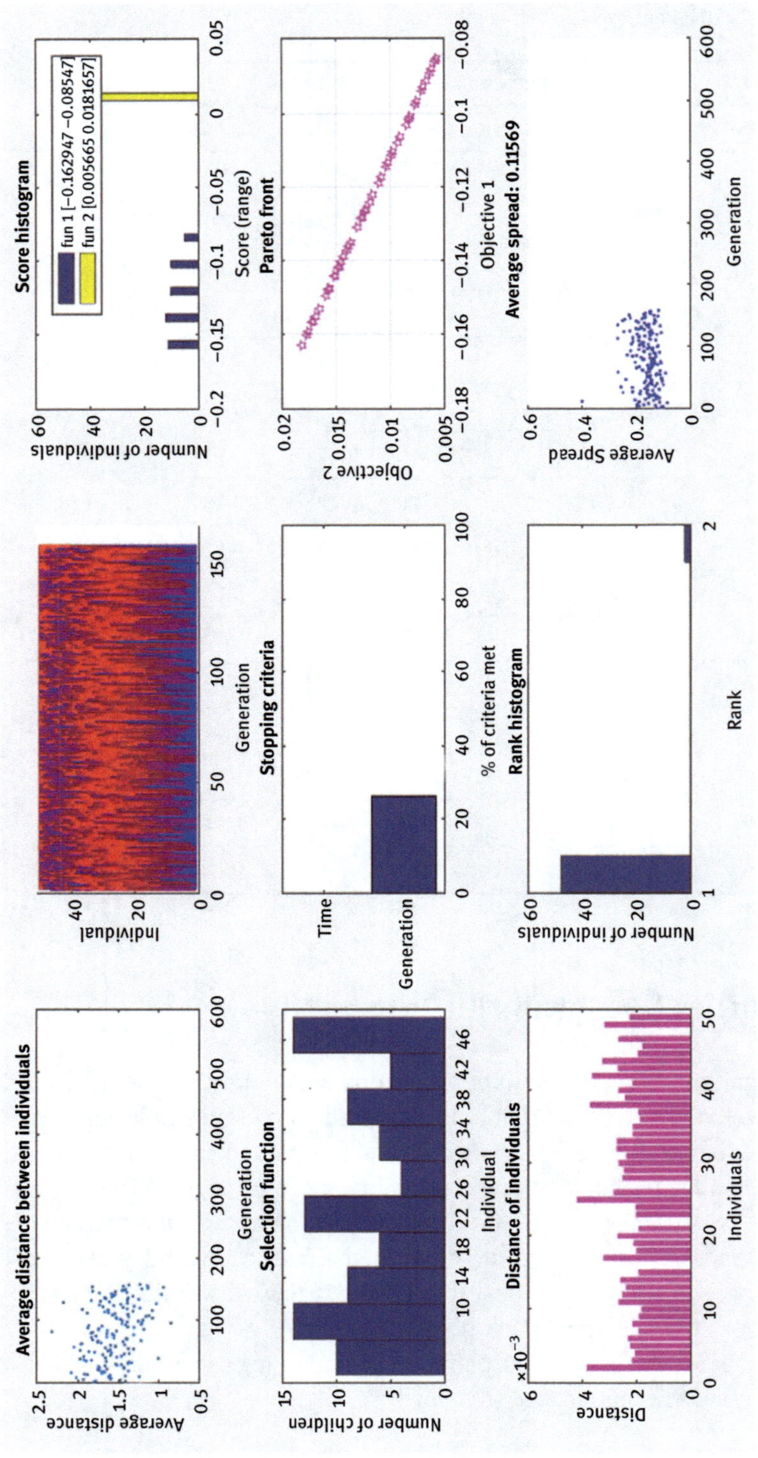

Figure 9.11: Graphical illustrations of GA.

$$X_{pq} = \frac{Y_{pq} - \mathrm{Min}[Y_{pq,}p = 1, 2, ..., n]}{\mathrm{Max}[Y_{pq,}p = 1, 2, ..., n] - \mathrm{Min}[Y_{pq,}p = 1, 2, ..., n]} \tag{9.12}$$

In order to obtain the minimized performance value for some parameter like EWR, then the equation is normalized as follows:

$$X_{pq} = \frac{\mathrm{Max}[Y_{pq,}i = 1, 2, ..., n] - Y_{pq}}{\mathrm{Max}[Y_{pq,}p = 1, 2, ..., n] - \mathrm{Min}[Y_{pq,}p = 1, 2, ..., n]} \tag{9.13}$$

The correlation among definite normalized data along with reference can be expressed by the gray relation coefficient as follows:

$$Y(X_{oq}, X_{pq}) = \frac{\nabla_{\min} + \varsigma\nabla_{\max}}{\nabla_{pq} + \varsigma\nabla_{\max}} \quad [p = 1, 2, ..., n \text{ and } q = 1, 2, ..., m] \tag{9.14}$$

where $\nabla_{ij} = |X_{oj} - X_{ij}|$, $\nabla_{\min} = \mathrm{Min}[\nabla_{pq}, p = 1, 2, ..., n \text{ and } q = 1, 2, ..., m]$ and $\nabla_{\max} = \mathrm{Max}[\nabla_{pq}, p = 1, 2, ..., n \text{ and } q = 1, 2, ..., m]$.

ς denotes the prominent coefficient (vary within 0–1).

In general, a suitable fraction can agree with the prominent coefficient as follows:

$$\Gamma(X_0, X_p) = \frac{1}{m}\sum_{p=1}^{m} Y(X_{0q}, X_{pq}) \tag{9.15}$$

where m denotes the quantity of process control parameter.

A single arrangement of process parameter can be achieved by the help of GRA. Here the parametric combination having rank 1 satisfies the multiresponse optimization.

9.7 Concept of fuzzy set

Whenever the problem of uncertainty in decision making arises, fuzzy set concept is applied. Instead of the numerical values, the linguistic calculation of weights is done here. The transformation of the decision matrix to the fuzzy decision matrix can be influenced by the decision makers' fuzzy ratings (Keufmann & Gupta, 1985). Finally, it creates the normalized fuzzy decision matrix.

An element X, which is a part of a fuzzy set $\hat{p}$ can be expressed by its membership function $\mu_{\hat{p}}(x)$, where "x" denotes the number of the membership and $\hat{d}$ maps all elements in X to the actual number in the limit of 0–1. At this point, triangular fuzzy number (TFN), $\hat{p}$, can be expressed as a triplet (p_1, p_2, p_3) and the membership function is defined [12] as follows:

$$\mu_{\tilde{p}}(x) = \begin{cases} 0, & x \le p_1 \\ \frac{x - p_1}{p_2 - p_1}, & p_1 \le x \le p_2 \\ \frac{p_3 - x}{p_3 - p_2}, & p_2 \le x \le p_3 \\ 0, & x > p_3 \end{cases} \tag{9.16}$$

The defuzzification is generation of a crisp value that relates to the formation of a nonfuzzy number from a fuzzy number

Here best nonfuzzy performance (BNP) number is calculated by using "centroid of area" technique:

$$\mathrm{BNP} = \frac{[(p_3 - p_1) - (p_2 - p_1)]}{3} + p_1 \tag{9.17}$$

Selecting the linguistic terms for criteria,

linguistics terms are utilized in order to give the weight of various criteria. By applying TFNs, the linguistic terms can be easily expressed as given in Table 9.3.

Table 9.3: Linguistic variables in case of criteria.

Linguistic variables	Fuzzy number
Very high (VH)	(0.9,1.0,1.0)
High (H)	(0.7,0.9,1.0)
Moderately high (MH)	(0.5,0.7,0.9)
Moderate (M)	(0.3,0.5,0.7)
Moderately low (ML)	(0.1,0.3,0.5)
Low (L)	(0.0,0.1,0.3)
Very low (VL)	(0.0,0.0, 0.1)

9.7.1 Fuzzy gray relational analysis

In this chapter, the two objective functions considered are MRR and EWR. They reflect different consequences while machining. During experimentation, more importance is imparted on EWR as compared to MRR [13]. While applying fuzzy logic, the weights of different criteria are shown in Table 9.4.

Table 9.4: Weights selected for the outputs.

Responses	Linguistic variables	Fuzzy number	BNP
MRR	MH	0.5,0.7,0.9	0.436
EWR	H	0.7,0.9,1	0.564

The highest gray relation grade value reflects that the result has reached the optimal condition. Therefore, the optimized machining condition is reached when the gray relation grade value is maximum. From the output of GA 17 number of results are obtained, which are basically the various experimental combinations. Now while applying MCDM, the conflicting nature of the objectives, that is, high MRR and low EWR, gets simultaneously satisfied.

9.7.2 Process control parameter optimization

Considering the respective weightage for MRR to be 43.6% and for EWR as 56.4%, respectively, in the fuzzy set, the gray relation coefficient is determined. The corresponding gray grade calculated is illustrated in Table 9.5.

Table 9.5: Gray relation coefficient and grades with corresponding ranks.

Experiment sequence	Objectives		Gray grade	Position
	MRR	EWR		
1	0.085	0.006	0.65	2
2	0.163	0.018	0.68	1
3	0.136	0.014	0.51	13
4	0.116	0.011	0.51	14
5	0.126	0.012	0.50	17
6	0.096	0.007	0.57	8
7	0.152	0.017	0.58	7
8	0.092	0.007	0.59	6
9	0.138	0.014	0.51	12
10	0.151	0.016	0.57	9
11	0.131	0.013	0.50	16
12	0.119	0.011	0.50	15
13	0.158	0.017	0.63	5
14	0.099	0.008	0.56	10
15	0.159	0.018	0.64	3
16	0.142	0.015	0.53	11
17	0.087	0.006	0.63	4

It is imperative from the table that the second experiment gives the maximum gray grade. So, it can be inferred that this experimental sequence satisfies multicriteria decision-making condition. Here the parametric combination of current (10 amp), voltage (45 volts), and pulse on time (52 µs) is best suited for obtaining high MRR and low EWR. Finally, validation experiment is conducted, while applying the above-mentioned experimental combination through gray relational grade the MRR is 0.163 and 0.0180 g/min, respectively. The comparison between predicted results and experimental results is shown in Table 9.6.

Table 9.6: Results of predicted performance experimental findings.

Criteria	Estimated result	Analytical result
MRR	0.163	0.162
EWR	0.0180	0.0182
Gray relation grade	0.68	0.60
Enhancement of the gray relation grade by **0.08**		

9.8 Conclusion

Nowadays, several nature-inspired metaheuristics help in solving complex optimization problems effectively. The multiobjective yet contradictory natured objective functions are easily optimized with these algorithms. The fundamental concepts of these algorithms are taken from nature. ANN can be effectively used to train as well as validate the experimental results. GA provides global optimal solutions to the multiobjective optimization problems. Finally, from the results obtained from the GA, a single parametric combination is derived using fuzzy multicriteria decision-making technique where contradictory objectives are satisfied simultaneously. Therefore, it is evident that EDM process can effectively machine HCHCr steel while varying current, voltage, and pulse on time, respectively, by using titanium nitride–coated copper tool. Through ANN, it is evident that the objective to maximize MRR and minimize EWR has been successfully accomplished since the value for R in both the cases is 1. From the GA algorithm, the values of MRR range between 0.085 and 0.163 g/min, and the values of EWR vary between 0.006 and 0.018 g/min, respectively. Considering the range of values of MRR and EWR, finally FGRA is applied. Lastly from the analysis, the parametric combination having current (10 amp), voltage (45 volts), and pulse on time (52 µs) provides the maximum MRR and minimum EWR simultaneously.

Now EDM is a very complex process having numerous control parameters. Today it has emerged as a key machining system in the manufacturing domain. Commercially, it is also proving to be very economical and productive. The ranges of these parameters are quite large. But through this experimental study, the exact value of the control parameters through which the effective machining can be done has been established. This will be useful to the operators who are working on it commercially.

References

[1] El-Hofy H. Advance machining process. McGraw Hill, New York, 2005.

[2] Kanthababu M. Multi-objective optimization of manufacturing processes using evolutionary algorithms. In: Computational methods for optimizing manufacturing technology: models and techniques, IGI Global, 2012, 44–66.

[3] Palanikumar K., Latha B., & Davim J.P. Application of Taguchi method with Grey Fuzzy Logic for the optimization of machining parameters in machining composites. In: Computational methods for optimizing manufacturing technology: models and techniques, IGI Global, 2012, 219–241.

[4] Vasant P. Novel meta-heuristic optimization techniques for solving fuzzy programming problems. In: Handbook of research on industrial informatics and manufacturing intelligence: Innovations and solutions, IGI Global, 2012, 104–131.

[5] Meyer D., Balemi A., & Wearing C. Neural networks – their use and abuse for small data sets. In: Heuristic and optimization for knowledge discovery, IGI Global, 2002, 169–185.

[6] Deo R.C., Ghimire S., Downs N.J., & Raj N. Optimization of wind speed prediction using an artificial neural network compared with a genetic programming model. In: Handbook of research on predictive modeling and optimization methods in science and engineering, IGI Global, 2018, 328–359.

[7] Sahu A.K. & Mahapatra S.S. Optimization of surface roughness parameters by different multi-response optimization techniques during electro-discharge machining of titanium alloy. In: Non-conventional machining in modern manufacturing systems, IGI Global, 2019, 82–108.

[8] Bose G.K., Pain P., & Roy S. Bio-inspired meta-heuristic multi-objective optimization of EDM process. In: Optimizing current strategies and applications in industrial engineering, IGI Global, 2019, 305–319.

[9] Pain P. & Bose G.K. Multi-objective optimization of ECG process applying soft computing techniques. In: Advanced fuzzy logic approaches in engineering science, IGI Global, 2018, 68–98.

[10] Prabhu S. & Kumar A. (2018) Investigation of nickel coated tool for electrochemical deburring of Al6082, International Journal of Surface Engineering and Interdisciplinary Materials Science (IJSEIMS), 1(6), 17–31.

[11] Majumder A. (2013) Process parameter optimization during EDM of AISI 316 LN stainless steel by using fuzzy based multi-objective PSO, Journal of Mechanical Science and Technology, 27 (7), 2143–2151.

[12] Dubois D. & Prade H. (1979) Operations in a fuzzy-valued logic, Information and Control, 43, 224–240.

[13] Neta B.M.M., Araújoa G.H.D., Guimarãesa F.G., Mesquitaa R.C., & Ekelc P.Y. (2012) A fuzzy genetic algorithm for automatic orthogonal graph drawing, Applied Soft Computing, 12(4), 1379–1389.

[14] Keufmann A. & Gupta M.M. Introduction to fuzzy arithmetic: theory and applications. Van Nostrand Reinhold, New York, 1985.

[15] Deng J. (1989) Introduction to grey system, The Journal of Grey System, 1(1), 1–24.

Pawan Kumar, S.B. Singh

10 Fuzzy reliability of system using different types of bifuzzy failure rates of components

Abstract: The present study proposes to determine the fuzzy reliability of systems using different types of conflicting bifuzzy numbers Till now, to evaluate the fuzzy reliability used the failure rates of components of systems follow the same type of fuzzy set or conflicting bifuzzy set. However, in real-life problems, such type of situation rarely occurs. Therefore, in this study, a new method has been introduced to determine the fuzzy reliability of a system having components following different types of conflicting bifuzzy failure rates. Using the introduced method, membership functions and nonmembership functions of fuzzy reliability of a series, parallel, parallel–series, and series–parallel systems are evaluated. Numerical problems are also taken to describe the proposed study.

Keywords: fuzzy system reliability, conflicting bifuzzy number, failure rate, system

10.1 Introduction

The traditional reliability of an item is the probability that item will perform a required function without failure under stated conditions for a stated period of time. More generally, reliability is the capacity of parts, components, equipments, products, and systems to perform their required functions for desired periods of time without failure, in specified environments and with a desired confidence. However, in real-life problems, the collected data or information are not always accurate hence the deduction of appropriate value of probability becomes very difficult in many cases. To handle this problem, fuzzy set theory has been used to analyze the reliability of a system. The fuzzy set theory introduced by Zadeh [1].

To understand the concept of fuzzy set, let us consider a system having three independent components. Evidently, the system is in working stage when all three components are in working condition simultaneously. However, when one or two components are failed, the system will operate in an offended state. In this stage, the system is neither fully working nor fully failed, but is in some intermediate state. This implies that the binary state assumption for delineate system failure and

Pawan Kumar, Department of Mathematics, Statistics and Computer Science
S.B. Singh, G.B. Pant University of Agriculture and Technology, Pantnagar, Uttarakhand, India

https://doi.org/10.1515/9783110628616-010

success may no longer be suitable. To handle this problem, we fuzzify the system failure. In real-life problems a person may assume that a commodity belongs to a set to a few degrees, but he is not sure that a commodity belongs to a set. In other words, there may be a hesitance or uncertainty about the membership function of object in A. An eventually solution is to use intuitionistic fuzzy sets (IFS), defined by Atanassov (1986).

The concept of IFS is the generalizations of the fuzzy set theory of Zadeh [1]. The membership function is considered only in fuzzy set but membership degree $\mu_A(x)$ and nonmembership function $v_A(x)$ are considered in IFS and the sum of membership and nonmembership degree is less than or equal to 1. Suppose if the sum of the value of membership and nonmembership function is greater than 1 then how do we fulfill the condition, for example, in a case where the membership function $\mu_A(x) = 0.8$ and the nonmembership function $v_A(x)$ equals to 0.6? This situation gives us a chance to redefine a solution lastly to reform the concept of IFS. Based on this reason Zamali et al. [2] gave a new concept of conflicting bifuzzy set (CBFS). In bifuzzy set, the sum of membership and nonmembership degree is less than 2. Hence, CBFS theory is an extension of IFS theory. Imran et al. [3] presented the definition and graphical representation of conflicting bifuzzy sets based on intuitionistic evaluation.

Fuzzy reliability theory was introduced and developed by Wen et al. [4, 5], Chen and Mon [6], Onisawa and Kacprzyk [7], and Cai [8]. The recent compilation of papers by Onisawa and Kacprzyk [7] gave several different approaches for fuzzy reliability. Kumar and Yadav [9] proposed a new method for analyzing fuzzy reliability of systems using different types of intuitionistic fuzzy failure rates of components. Aliev et al. [10] analyzed fuzzy reliability of systems using time dependent fuzzy set of components. Mon and Cheng [11] developed a new method for fuzzy reliability of systems for components with different membership functions. Kumar et al. [12] developed a new method for evaluating fuzzy reliability using intuitionistic fuzzy number. Mon et al. [11] analyzed the fuzzy reliability of different systems with different membership functions.

In the above-mentioned studies, it is supposed that the failure rates of components of systems follow the conflicting bifuzzy numbers (CBFNs) to evaluate the fuzzy reliability of systems. In this study, we have constructed the membership degree and nonmembership degree of the fuzzy system reliability using different types of CBFNs. Using introduced method, membership degree and non-membership degree of fuzzy system reliability of series, parallel, parallel–series, and series–parallel systems are evaluated. Numerical problems are also taken to exemplify the proposed study.

10.2 Preliminaries

Definition 10.1 ([1]). If a set Y be fixed, then a fuzzy set $\tilde{Z}$ is given by

$$\tilde{Z} = \{\langle y, \mu_{\tilde{Z}}(y)\rangle : y \in Y\}$$

where $\mu_{\tilde{Z}}(x) \in [0, 1]$ is the membership function of the element $y \in$ Y.

Definition 10.2 (Atanassov 1986). If a set Z be fixed then an intuitionistic fuzzy set (IFS) $\tilde{A}$ is given by

$$\tilde{A} = \{\langle z, \mu_{\tilde{A}}(z), \nu_{\tilde{A}}(z) : z \in Z\rangle\}$$

where $\mu_{\tilde{A}}: Z \to [0, 1]$ is the membership function of positive $z, (z \in Z)$ and $\nu_{\tilde{A}}: Z \to [0, 1]$ is the non- membership function of negative $z, (z \in Z)$ w.r.t. $\tilde{A}$ and $0 \le \mu_{\tilde{A}} + \nu_{\tilde{A}} \le 1$.

Zamali [2] in his study proposed the new notion called a bifuzzy set which is extension of IFS. He defines CBFS.

Definition 10.3 ([2]). If a set A is fixed then a CBFS $\tilde{Z}$ of A is an object having the given form as follows

$$\tilde{Z} = \{\langle a, \mu_{\tilde{Z}}(a), \nu_{\tilde{Z}}(a) : a \in A\rangle\}$$

where $\mu_{\tilde{Z}}: A \to [0, 1]$ is the membership function of positive a, $(a \in A)$ with respect to $\tilde{Z}$ and $\nu_{\tilde{Z}}: A \to [0, 1]$ is the nonmembership function of negative $a, (a \in A)$ with respect to $\tilde{Z}$ and $0 \le \mu_{\tilde{Z}}(a) + \nu_{\tilde{Z}}(a) \le 2$.

10.2.1 α-Level set or α-Cut set of a CBFS

α-Cut of a CBFS $\tilde{A}$ is defined as

$$\tilde{A}_\alpha = \{x \in X : \mu_{\tilde{A}}(x) \ge \alpha, \}; 0 \le \alpha$$

10.2.2 β-Level set or β-Cut set of a CBFS

β-Cut of a CBFS $\tilde{A}$ is defined as

$$\tilde{A}_\beta = \{x \in X : \nu_{\tilde{A}}(x) \le \beta\}; \beta \le 1$$

10.2.3 Triangular CBFS

A triangular CBFS $\tilde{A}$ is defined as

$$\tilde{A} = (m - l, m, m + n; m - l', m, m + n')$$

where $m \in R$ is the center, $l > 0$ and $n > 0$ are the left and right dispersion of the membership function of $\tilde{A}$, respectively. $l' > 0$ and $n' > 0$ are the left and right spreads of the nonmembership function of $\tilde{A}$, respectively.

10.2.4 Trapezoidal CBFS

$$\tilde{A} = (p - l, q, r, s + n; p - l', q, r, s + n')$$

where $p, q, r, s \in R$ is the center, $l > 0$ and $n > 0$ are the left and right dispersion of the $\mu_{\tilde{A}}(x)$, respectively. $l' > 0$ and $n' > 0$ are the left and right dispersion of the $v_{\tilde{A}}(x)$, respectively.

10.3 Bifuzzy reliability function or survival function

The reliability function R (t) of systems in terms of failure rate λ (t) is given by following formula

$$R(t) = \exp\left[-\int_0^t \lambda(t')dt'\right], \qquad t > 0 \tag{10.1}$$

Let us consider the failure rate is constant, that is, $(t) = \lambda$, then the reliability is as follows:

$$R(t) = \exp(-\lambda t) \tag{10.2}$$

If failure rate is not precise, then failure rate is considered as CBFS.

Let failure rate is considered as CBFS $\tilde{\lambda}$ defined on Z given below:

$$\tilde{\lambda} = \{ < z, \mu_{\tilde{\lambda}}(z), v_{\tilde{\lambda}}(z) > : z \in Z\}$$

Two types of cuts for CBFS are defined as follows:

α-Cut of membership function $\mu_{\tilde{\lambda}}(z)$ which is defined as

$$\tilde{\lambda}_{\alpha} = \{z : \mu_{\tilde{\lambda}}(z) \geq \alpha, \alpha \in [0, 1]\}$$

β-Cut of nonmembership function $v_{\tilde{\lambda}}(z)$ is as

$$\tilde{\lambda}_\beta = \{z: v_{\tilde{\lambda}}(z) \le \beta, \beta \in [0,1]\}$$

Failure rate $\tilde{\lambda}$ is in form of CBFS. Then fuzzy-convexity of CBFS is as follows:

$$\tilde{\lambda}_\alpha = [\lambda_\alpha^L, \lambda_\alpha^R] \quad \forall \alpha \in [0,1]$$

Also fuzzy-concavity of CBFS is as follows:

$$\tilde{\lambda}_\beta = [\lambda_\beta^L, \lambda_\beta^R] \quad \forall \beta \in [0,1]$$

where λ_α^L and λ_α^R increasing and decreasing function of α and β, respectively. λ_β^L and λ_β^R increasing and decreasing function of α and β, respectively, with $\alpha, \beta \in [0,1]$.

The reliability function $\bar{R}(t)$ using CBFS defined on X is as follows:

$$\bar{R}(t) = \{ < x, \mu_{\bar{R}(t)}(x), v_{\bar{R}(t)}(x) > : x \in X\}$$

Now failure rate follows CBFS, then the reliability $\bar{R}(t)$ is also in the form of CBFS.

Now α-Cut for membership function of $\bar{R}(t)$ is

$$\bar{R}_\alpha(t) = [r_\alpha^L(t), r_\alpha^R(t)]$$

where

$$r_\alpha^L(t) = \min(e^{-xt}) \quad \text{s.t. } \lambda_\alpha^L \le x \le \lambda_\alpha^R \tag{10.3a}$$

$$r_\alpha^R(t) = \max(e^{-xt}) \quad \text{s.t. } \lambda_\alpha^L \le x \le \lambda_\alpha^R \tag{10.3b}$$

β-Cut for nonmembership function of $\bar{R}(t)$ is given by

$$\bar{R}_\beta(t) = [r_\beta^L(t), r_\beta^R(t)]$$

where

$$r_\beta^L(t) = \min(e^{-xt}) \quad \text{s.t. } \lambda_\beta^L \le x \le \lambda_\beta^R \tag{10.4a}$$

$$r_\beta^R(t) = \max(e^{-xt}) \quad \text{s.t. } \lambda_\beta^L \le x \le \lambda_\beta^R \tag{10.4b}$$

For membership and nonmembership function, α-Cut and β-Cut of $\tilde{R}(t)$ respectively, is

$$\bar{R}_\alpha(t) = [e^{-\lambda_\alpha^R.t}, e^{-\lambda_\alpha^L.t}] \tag{10.5a}$$

$$\bar{R}_\beta(t) = [e^{-\lambda_\beta^R.t}, e^{-\lambda_\beta^L.t}] \tag{10.5b}$$

10.4 Fuzzy reliability of series system using different types of conflicting bifuzzy failure rates

Consider a system having n components connected in series (Figure 10.1). Let failure rates of n components are represented by different CBFN $\tilde{\lambda}_1, \tilde{\lambda}_2, ..., \tilde{\lambda}_n$ and reliability with conflicting bifuzzy failure rates $\tilde{\lambda}_1, \tilde{\lambda}_2, ..., \tilde{\lambda}_n$ are $\tilde{R}_1, \tilde{R}_2, ..., \tilde{R}_n$, respectively, at time t.

Figure 10.1: Series system.

Reliability of series system at time t is

$$\bar{R}_S(t) = \bar{R}_1(t) \otimes \bar{R}_2(t) \otimes \cdots \otimes \bar{R}_n(t) \tag{10.6}$$

From eq. (10.2), we get

$$\bar{R}_i(t) = e^{-\bar{\lambda}_i . t}, \ i = 1, 2, 3, \ldots, n; \ t > 0 \tag{10.7}$$

The reliability of series system $\bar{R}_s(t)$ is estimated as

$$\bar{R}_s(t) = e^{-t \sum_{J=1}^{n} \bar{\lambda}_J} = e^{-\bar{\lambda}_s . t} \tag{10.8}$$

where $\bar{\lambda}_s = \psi(\bar{\lambda}_1, \bar{\lambda}_2, ..., \bar{\lambda}_n) = \Sigma_{j=1}^{n} \bar{\lambda}_j$ is failure rate.

α-Cut of $\bar{\lambda}_s$ for membership function is

$$\tilde{\lambda}_{j\alpha} = [h_{j\alpha}^L, h_{j\alpha}^R] \quad \forall \alpha \in [0, 1], \ j = 1, 2, \ldots, n$$

We have n intervals $\tilde{\lambda}_{1\alpha} = [h_{1\alpha}^L, h_{1\alpha}^R], \tilde{\lambda}_{2\alpha} = [h_{2\alpha}^L, h_{2\alpha}^R], \ \ldots, \ \tilde{\lambda}_{1\alpha} = [h_{n\alpha}^L, h_{n\alpha}^R] \ \forall \tilde{\lambda}_j, \ 1 \leq j \leq n$, respectively, for every α.

β-Cut of nonmembership function for $\tilde{\lambda}_S$ is

$$\tilde{\lambda}_{j\beta} = [h_{j\beta}^L, h_{j\beta}^R] \quad \forall \beta \in [0, 1], \ j = 1, 2, \ldots, n$$

Similarly, we have n intervals $\tilde{\lambda}_{1\beta} = [h_{1\beta}^L, h_{1\beta}^R], \ \tilde{\lambda}_{2\beta} = [h_{2\beta}^L, h_{2\beta}^R], \ \ldots, \ \tilde{\lambda}_{1\beta} = [h_{n\beta}^L, h_{n\beta}^R] \ \forall \tilde{\lambda}_j$, $1 \leq j \leq n$, respectively, for every β.

α-Cut of $\tilde{\lambda}_S$ for membership function is

$$\tilde{\lambda}_{S\alpha} = [\psi_{S\alpha}^L, \psi_{S\alpha}^R] \quad \forall \alpha \in [0, 1]$$

where

$$\psi^L_{S\alpha} = \min \sum_{j=1}^{n} x_j,\ \ \psi^R_{S\alpha} = \max \sum_{j=1}^{n} x_j$$

$$s.t.\ h^L_{1\alpha} \le x_1 \le h^R_{1\alpha},\ h^L_{2\alpha} \le x_2 \le h^R_{2\alpha}, \dots,\ h^L_{n\alpha} \le x_n \le h^R_{n\alpha} \tag{10.9}$$

β-Cut of $\bar{\lambda}_s$ for nonmembership function is

$$\tilde{\lambda}_{S\beta} = [\psi^L_{S\beta}, \psi^R_{S\beta}] \quad \forall \beta \in [0,1]$$

where

$$\psi^L_{S\beta} = \min \sum_{j=1}^{n} x_j,\ \psi^R_{S\beta} = \max \sum_{j=1}^{n} x_j$$

$$\text{s.t. } h^L_{1\beta} \le x_1 \le h^R_{1\beta}, h^L_{1\beta} \le x_1 \le h^R_{1\beta}, \dots, h^L_{1\beta} \le x_1 \le h^R_{1\beta} \tag{10.10}$$

After solving (10.9) and (10.10), we find α-Cut and β-Cut of fuzzy system reliability.

α-Cut and β-Cut of membership function of $\bar{R}_s(t)$ as

$$\tilde{R}_{S\alpha}(t) = [e^{-\psi^R_{S\alpha}.t}, e^{-\psi^L_{S\alpha}.t}] \tag{10.11a}$$

$$\tilde{R}_{S\beta}(t) = [e^{-\psi^R_{S\beta}.t}, e^{-\psi^L_{S\beta}.t}] \tag{10.11b}$$

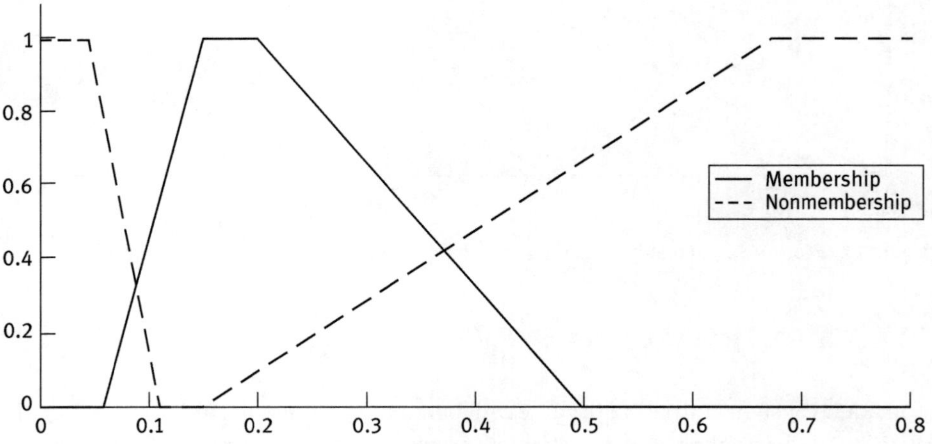

Figure 10.2: For $t = 10$, membership and nonmembership function of reliability of radio set.

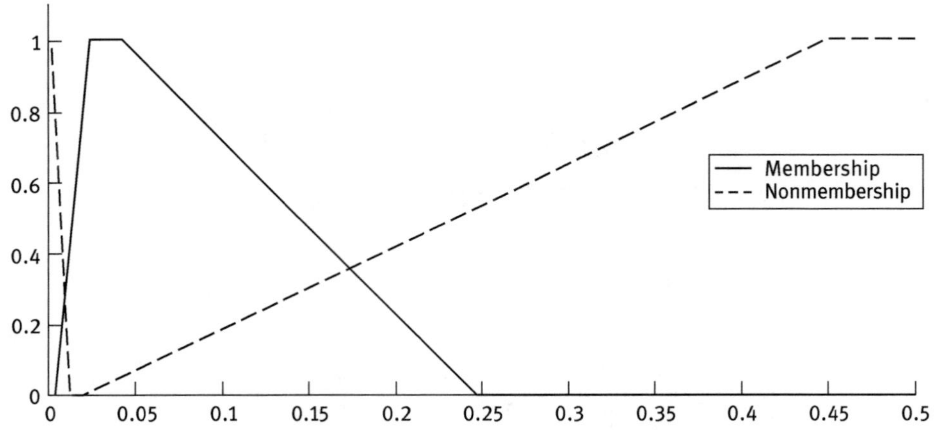

Figure 10.3: For $t = 20$, membership and nonmembership function of reliability of radio set.

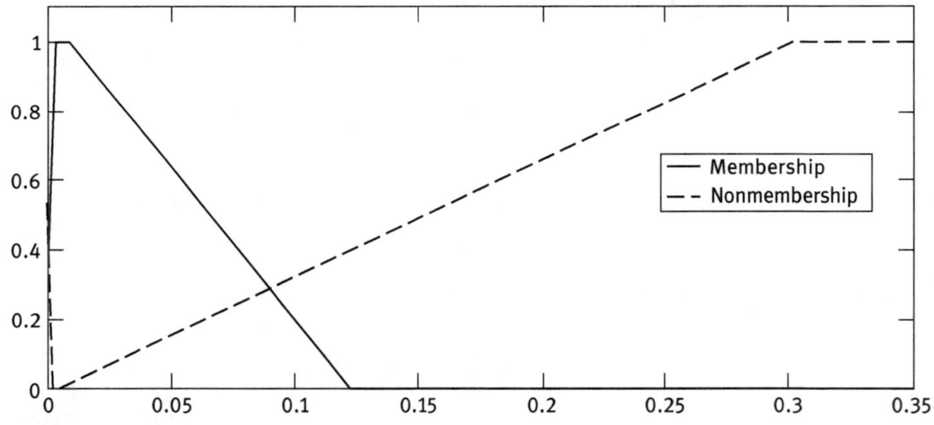

Figure 10.4: For $t = 30$, membership and nonmembership function of reliability of radio set.

10.4.1 Numerical problem

Consider a system consisting of three independent components and all components work properly to operate the system successfully. Let failure rates of components are in the form of different types of CBFS $\tilde{\lambda}_1$, $\tilde{\lambda}_2$, and $\tilde{\lambda}_3$, respectively, assumed values of $\tilde{\lambda}_1$, $\tilde{\lambda}_2$, $\tilde{\lambda}_3$ are given in Table 10.1.

Table 10.1: Conflicting bifuzzy failure rates of components.

Bifuzzy failure rate of *i*th component	Types of bifuzzy number
$\tilde{\lambda}_1 = (0.02,\ 0.05,\ 0.08;\ 0.01,\ 0.05,\ 0.09)$	Triangular conflicting bifuzzy number
$\tilde{\lambda}_2 = (0.03,\ 0.06,\ 0.08;\ 0.02,\ 0.07,\ 0.09)$	Triangular conflicting bifuzzy number
$\tilde{\lambda}_3 = (0.02,\ 0.05,\ 0.08,\ 0.12;\ 0.01,\ 0.07,\ 0.10,\ 0.13)$	Trapezoidal conflicting bifuzzy number

α-Cuts of $\tilde{\lambda}_1$, $\tilde{\lambda}_2$, $\tilde{\lambda}_3$ are estimated and tabulated in Table 10.2.

Table 10.2: α-Cut of $\tilde{\lambda}_1$, $\tilde{\lambda}_2$ and $\tilde{\lambda}_3$ for membership.

α	$\tilde{\lambda}_{1\alpha}$	$\tilde{\lambda}_{2\alpha}$	$\tilde{\lambda}_{3\alpha}$
0.0	[0.02,0.08]	[0.03, 0.08]	[0.02, 0.12]
0.1	[0.023,0.077]	[0.033, 0.078]	[0.023, 0.116]
0.2	[0.026,0.074]	[0.036, 0.076]	[0.026, 0.112]
0.3	[0.029,0.071]	[0.039, 0.074]	[0.029, 0.108]
0.4	[0.032,0.068]	[0.042, 0.072]	[0.032, 0.104]
0.5	[0.035,0.065]	[0.045, 0.070]	[0.035, 0.100]
0.6	[0.038,0.062]	[0.048, 0.068]	[0.038, 0.096]
0.7	[0.041,0.059]	[0.051, 0.066]	[0.041, 0.092]
0.8	[0.044,0.056]	[0.054, 0.064]	[0.044, 0.088]
0.9	[0.047,0.053]	[0.057, 0.062]	[0.047, 0.084]
1.0	[0.05, 0.05]	[0.06, 0.06]	[0.05, 0.08]

β-Cuts of $\tilde{\lambda}_1$, $\tilde{\lambda}_2$, $\tilde{\lambda}_3$ are estimated and tabulated in Table 10.3.

Table 10.3: β-Cut of $\tilde{\lambda}_1$, $\tilde{\lambda}_2$, and $\tilde{\lambda}_3$ for nonmembership.

β	$\tilde{\lambda}_{1\beta}$	$\tilde{\lambda}_{2\beta}$	$\tilde{\lambda}_{3\beta}$
0.0	[0.05, 0.05]	[0.07, 0.07]	[0.07, 0.10]
0.1	[0.046, 0.054]	[0.065, 0.072]	[0.064, 0.103]
0.2	[0.042, 0.058]	[0.060, 0.074]	[0.058, 0.106]
0.3	[0.038, 0.062]	[0.055, 0.076]	[0.052, 0.109]
0.4	[0.034, 0.066]	[0.050, 0.078]	[0.046, 0.112]

Table 10.3 (continued)

β	$\tilde{\lambda}_{1\beta}$	$\tilde{\lambda}_{2\beta}$	$\tilde{\lambda}_{3\beta}$
0.5	[0.03, 0.07]	[0.045, 0.080]	[0.040, 0.115]
0.6	[0.026, 0.074]	[0.040, 0.082]	[0.034, 0.118]
0.7	[0.022, 0.078]	[0.035, 0.084]	[0.028, 0.121]
0.8	[0.018, 0.082]	[0.030, 0.086]	[0.022, 0.124]
0.9	[0.014, 0.086]	[0.025, 0.088]	[0.016, 0.127]
1.0	[0.01, 0.09]	[0.02, 0.09]	[0.01, 0.13]

Using formulas (10.9) to (10.11), we get fuzzy reliability of series system

$$\tilde{R}_{S\alpha}(t) = [e^{-(.28-.09\alpha)t}, e^{-(.07+.09\alpha)t}] \tag{10.12a}$$

$$\tilde{R}_{S\beta}(t) = [e^{-(.13+.09\beta)t}, e^{-(.19-.15\beta)t}] \tag{10.12b}$$

The fuzzy system reliability of given system is calculated by eq. (10.12). α-Cut and β-Cut of fuzzy reliability are computed for $t = 10$, 20, 30 shown in Table 10.4 and fig. 10.2, 10.3 & 10.4. The fuzzy reliability of above system is in trapezoidal CBFN for failure rates as conflicting bifuzzy of components for different time t.

10.5 Fuzzy reliability of parallel system using of conflicting bifuzzy failure rates

Consider a system consisting "n" components are in parallel form (Figure 10.5). Let us consider failure rates of n components are in form of different CBFNs $\tilde{\lambda}_1, \tilde{\lambda}_2, \ldots, \tilde{\lambda}_n$. Let the fuzzy reliability of n components with conflicting bifuzzy failure rates are $\tilde{\lambda}_1, \tilde{\lambda}_2, \ldots, \tilde{\lambda}_n$ and $\tilde{R}_1, \tilde{R}_2, \ldots, \tilde{R}_n$, respectively, at time t.

Reliability of parallel system at time t is

$$\tilde{R}_P(t) = 1 - \prod_{i=1}^{n}(1 - \tilde{R}_i(t)) = 1 - \prod_{i=1}^{n}(1 - \exp(-\tilde{\lambda}_i.t)) \tag{10.13}$$

So $\tilde{R}_P(t) = \psi(\tilde{\lambda}_1, \tilde{\lambda}_2, \ldots, \tilde{\lambda}_n; t) = 1 - \prod_{i=1}^{n}(1 - \exp(-\tilde{\lambda}_i.t))$

α-Cut of $\tilde{\lambda}_i$ for membership function is

$$\tilde{\lambda}_{i\alpha} = [h_{i\alpha}^L, h_{i\alpha}^R] \quad \forall \alpha \in [0, 1],\ i = 1, 2, 3, \ldots, n$$

We have n intervals $\tilde{\lambda}_{1\alpha} = [h_{1\alpha}^L, h_{1\alpha}^R]$, $\tilde{\lambda}_{2\alpha} = [h_{2\alpha}^L, h_{2\alpha}^R]$, $\ldots$, $\tilde{\lambda}_{n\alpha} = [h_{n\alpha}^L, h_{n\alpha}^R]$ $\forall \tilde{\lambda}_j$, $1 \le j \le n$ respectively, for each α.

Table 10.4: α-Cut and β-Cut of series system reliability $\bar{R}_s(t)$ for $t = 10, 20, 30$.

α, β	For $t = 10$		For $t = 20$		For $t = 30$	
	α-Cut	β-Cut	α-Cut	β-Cut	α-Cut	β-Cut
0	[.0608, .4966]	[.1108, .1496]	[.00369, .24659]	[.01228, .02237]	[.000225, .122456]	[.001360, .003346]
1	[.0665, .4538]	[.1013, .1737]	[.00427, .20598]	[.01026, .03019]	[.000295, .093481]	[.001040, .005247]
2	[.0728, .4148]	[.0926, .2018]	[.00530, .17204]	[.00856, .04076]	[.000386, .071361]	[.000790, .008229]
3	[.0797, .3791]	[.0846, .2345]	[.00635, .14370]	[.00715, .05502]	[.000505, .054476]	[.000610, .012907]
4	[.0872, .3465]	[.0773, .2725]	[.00759, .12003]	[.00598, .07427]	[.000662, .041586]	[.000460, .020242]
5	[.0954, .3166]	[.0707, .3166]	[.00909, .10026]	[.00499, .10026]	[.000867, .031746]	[.000350, .031746]
6	[.1044, .2894]	[.0646, .3679]	[.01088, .08374]	[.00417, .13534]	[.001136, .024234]	[.000270, .049787]
7	[.1142, .2645]	[.0590, .4274]	[.01304, .06995]	[.00348, .18268]	[.001488, .018499]	[.000210, .078082]
8	[.1249, .2417]	[.0539, .4966]	[.01561, .05843]	[.00291, .24659]	[.001949, .014122]	[.000156, .122456]
9	[.1367, .2209]	[.0493, .5769]	[.01868, .04880]	[.00243, .33287]	[.002554, .010781]	[.000119, .192049]
1	[.1496, .2018]	[.0451, .6703]	[.02237, .04076]	[.00203, .44933]	[.003346, .008229]	[.000092, .301194]

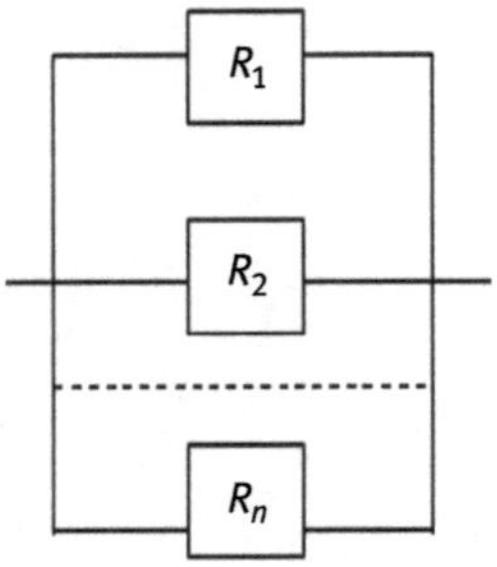

Figure 10.5: Parallel system.

β-Cut of nonmembership function of $\tilde{\lambda}_i$ is

$$\tilde{\lambda}_{i\beta} = [h^L_{i\beta}, h^R_{i\beta}] \quad \forall \beta \in [0,1], \ i = 1, 2, 3, \ldots, n.$$

Similarly, we have n intervals $\tilde{\lambda}_{1\beta} = [h^L_{1\beta}, h^R_{1\beta}]$, $\tilde{\lambda}_{2\beta} = [h^L_{2\beta}, h^R_{2\beta}]$, $\ldots$, $\tilde{\lambda}_{n\beta} = [h^L_{n\beta}, h^R_{n\beta}]$ $\forall \tilde{\lambda}_j$, $1 \le j \le n$, respectively, for each β.

α-Cut of $\tilde{R}_P(t)$ for membership function is

$$\tilde{R}_{P\alpha} = [\psi^L_{P\alpha}(t), \psi^R_{P\alpha}(t)]$$

where

$$\psi^L_{P\alpha}(t) = \min\left(1 - \prod_{j=1}^{n} (1 - e^{-x_j.t})\right), \ \psi^L_{P\alpha}(t) = \max\left(1 - \prod_{j=1}^{n} (1 - e^{-x_j.t})\right) \tag{10.14}$$

s.t. $h^L_{1\alpha} \le x_1 \le h^R_{1\alpha}, h^L_{2\alpha} \le x_2 \le h^R_{2\alpha}, \ldots, h^L_{n\alpha} \le x_n \le h^R_{n\alpha}$.

β-Cut of nonmembership function of $\bar{R}_P(t)$ is as

$$\tilde{R}_{P\beta} = [\psi^L_{P\beta}(t), \psi^R_{P\beta}(t)]$$

where

$$\psi^L_{P\beta}(t) = \min\left(1 - \prod_{i=1}^{n} (1 - e^{-x_i.t})\right), \ \psi^R_{P\beta}(t) = \max\left(1 - \prod_{i=1}^{n} (1 - e^{-x_i.t})\right) \tag{10.15}$$

s.t. $h^L_{1\beta} \le x_1 \le h^R_{1\beta}, h^L_{1\beta} \le x_1 \le h^R_{1\beta}, \ldots, h^L_{1\beta} \le x_1 \le h^R_{1\beta}$.

After solving eqs. (10.14) and (10.15), we find α-Cut and β-Cut of fuzzy reliability of parallel system $\bar{R}_P(t)$.

10.5.1 Numerical problem

Consider a system has three independent components. At least one of them must work properly to operate given system properly. Let conflicting bifuzzy failure rates of components are in the form of CBFNs $\tilde{\lambda}_1$, $\tilde{\lambda}_2$, and $\tilde{\lambda}_3$, respectively. For numerical computation assumed values of $\tilde{\lambda}_1$, $\tilde{\lambda}_2$, and $\tilde{\lambda}_3$ are given in Table 10.5.

Table 10.5: Conflicting bifuzzy failure rates of components.

Bifuzzy failure rate	Types of bifuzzy number
$\tilde{\lambda}_1 = (0.3, 0.4, 0.5; 0.2, 0.4, 0.6)$	Triangular conflicting bifuzzy number
$\tilde{\lambda}_2 = (0.2, 0.5, 0.7; 0.3, 0.6, 0.8)$	Triangular conflicting bifuzzy number
$\tilde{\lambda}_3 = (0.03, 0.05, 0.08, 0.10; 0.02, 0.06, 0.10, 0.12)$	Trapezoidal conflicting bifuzzy number

α-Cuts of $\tilde{\lambda}_1$, $\tilde{\lambda}_2$, $\tilde{\lambda}_3$ are evaluated and given in Table 10.6.

Table 10.6: For membership, α-Cut of $\tilde{\lambda}_1$, $\tilde{\lambda}_2$, and $\tilde{\lambda}_3$.

α	$\tilde{\lambda}_{1\alpha}$	$\tilde{\lambda}_{2\alpha}$	$\tilde{\lambda}_{3\alpha}$
0.0	[0.3, 0.5]	[0.2, 0.7]	[0.03, 0.10]
0.1	[0.31, 0.49]	[0.23, 0.68]	[0.032, 0.098]
0.2	[0.32, 0.48]	[0.26, 0.66]	[0.034, 0.096]
0.3	[0.33, 0.47]	[0.29, 0.64]	[0.036, 0.094]
0.4	[0.34, 0.46]	[0.32, 0.62]	[0.038, 0.092]
0.5	[0.35, 0.45]	[0.35, 0.60]	[0.04, 0.09]
0.6	[0.36, 0.44]	[0.38, 0.58]	[0.042, 0.088]
0.7	[0.37, 0.43]	[0.41, 0.56]	[0.044, 0.086]
0.8	[0.38, 0.42]	[0.44, 0.54]	[0.046, 0.084]
0.9	[0.39, 0.41]	[0.47, 0.52]	[0.048, 0.082]
1.0	[0.4, 0.4]	[0.5, 0.5]	[0.05, 0.08]

β-Cut of $\tilde{\lambda}_1$, $\tilde{\lambda}_2$, and $\tilde{\lambda}_3$ are evaluated and given in Table 10.7.

Table 10.7: For nonmembership, β-Cut of $\tilde{\lambda}_1$, $\tilde{\lambda}_2$, and $\tilde{\lambda}_3$.

β	$\tilde{\lambda}_{1\beta}$	$\tilde{\lambda}_{2\beta}$	$\tilde{\lambda}_{3\beta}$
0.0	[0.4, 0.4]	[0.6, 0.6]	[0.06, 0.10]
0.1	[0.38, 0.42]	[0.57, 0.62]	[0.056, 0.102]
0.2	[0.36, 0.44]	[0.54, 0.64]	[0.052, 0.104]
0.3	[0.34, 0.46]	[0.51, 0.66]	[0.048, 0.106]
0.4	[0.32, 0.48]	[0.48, 0.68]	[0.044, 0.108]
0.5	[0.3, 0.5]	[0.45, 0.70]	[0.04, 0.11]
0.6	[0.28, 0.52]	[0.42, 0.72]	[0.036, 0.112]
0.7	[0.26, 0.54]	[0.39, 0.74]	[0.032, 0.114]
0.8	[0.24, 0.56]	[0.36, 0.76]	[0.028, 0.116]
0.9	[0.22, 0.58]	[0.33, 0.78]	[0.024, 0.118]
1.0	[0.2, 0.6]	[0.3, 0.8]	[0.02, 0.12]

Using formulas (10.14) to (10.15), we get

$$\begin{aligned}
\psi^L_{P\alpha}(t) &= 1-(1-e^{-(.5-.1\alpha)t})\times(1-e^{-(.7-.2\alpha)t})\times(1-e^{-(.1-.02\alpha)t})\\
\psi^R_{P\alpha}(t) &= 1-(1-e^{-(.3+.1\alpha)t})\times(1-e^{-(.2+.3\alpha)t})\times(1-e^{-(.03+.02\alpha)t})\\
\psi^L_{P\beta}(t) &= 1-(1-e^{-(.4+.2\beta)t})\times(1-e^{-(.6+.2\beta)t})\times(1-e^{-(.1+.02\beta)t})\\
\psi^R_{P\beta}(t) &= 1-(1-e^{-(.4-.2\beta)t})\times(1-e^{-(.6-.3\beta)t})\times(1-e^{-(.06-.04\beta)t})
\end{aligned} \tag{10.16}$$

The system reliability of given system is evaluated using (10.16). α-Cut and β-Cut of fuzzy reliability of given parallel system are computed for $t = 10$, 20, 30 and are tabulated in Table 10.8 and Fig. 10.6, 10.7 & 10.8. The fuzzy reliability of system is in trapezoidal CBFN.

10.6 Reliability of parallel–series system when failure rates follow conflicting bifuzzy numbers

Let us consider a parallel–series system consisting "j" branches connected in parallel form and each branch contains "" components connected in series form as

Table 10.8: α-Cut and β-Cut of parallel system reliability $\tilde{R}_P(t)$ for $t = 10, 20, 30$.

α, β	For $t = 10$	For $t = 20$	For $t = 30$
	α-Cut β-Cut	α-Cut β-Cut	α-Cut β-Cut
0	[.37271, .78705] [.38099, .55817]	[.13538, .55817] [.13563, .30143]	[.04978, .40811] [.04979, .16530]
1	[.38065, .76470] [.37146, .58220]	[.14091, .53299] [.13023, .32662]	[.05286, .38357] [.04689, .18638]
2	[.38880, .74405] [.36245, .60738]	[.14667, .51015] [.12506, .35395]	[.05614, .36090] [.04416, .21015]
3	[.39717, .72485] [.35391, .63375]	[.15266, .489] [.12012, .38360]	[.05961, .33974] [.04159, .23696]
4	[.40577, .7068] [.34576, .66136]	[.15891, .46914] [.11539, .41579]	[.06329, .31989] [.03916, .26718]
5	[.41462, .68993] [.33797, .69021]	[.16541, .45033] [,11084, .45076]	[.06721, .30123] [.03688, .30128]
6	[.42372, .67388] [.33049, .72032]	[.17218, .43242] [.10649, .48876]	[.07136, .28368] [.03474, .33975]
7	[.43309, .65859] [.32330, .75162]	[.17923, .41530] [.10230, .52009]	[.07578, .26715] [.03271, .38315]
8	[.44276, .64396] [.31637, .78401]	[.18657, .39891] [.09829, .57506]	[.08046, .25159] [.03081, .43215]
9	[.45274, .62989] [.30966, .81727]	[.19423, .38319] [.09443, .62398]	[.08544, .23693] [.02901, .48748]
1	[.46306, .61634] [.30316, .85107]	[.20220, .36812] [.09072, .67716]	[.09072, .22314] [.02732, .54998]

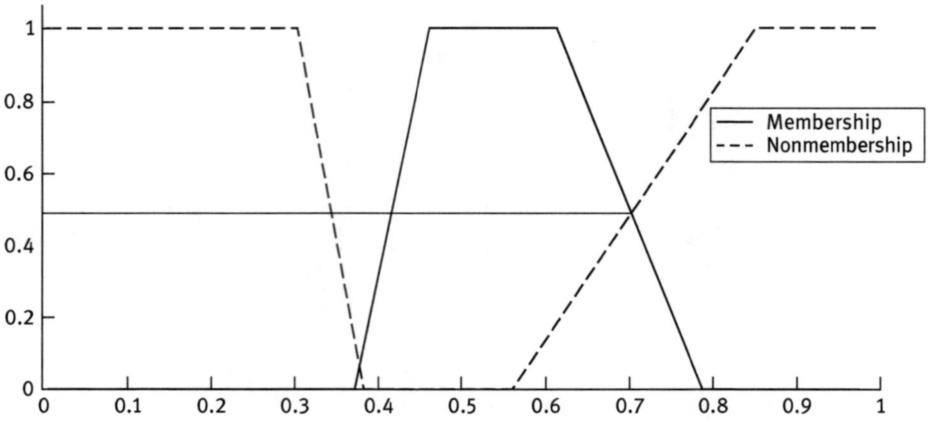

Figure 10.6: For $t = 10$, membership function and nonmembership function of reliability of achieving orbit.

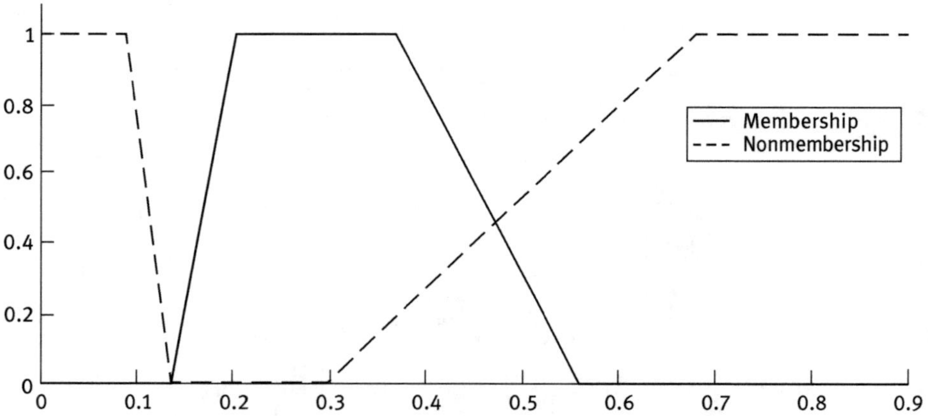

Figure 10.7: For $t = 20$, membership function and nonmembership function of reliability of achieving orbit.

shown in Figure 10.9. Let us consider failure rates of all components are following different CBFNs $\tilde{\lambda}_{rs}$.

Let failure rate function of the sth component of rth branch ($r = 1, 2, \ldots, i$ and $s = 1, 2, \ldots, j$) is represented by CBFS $\tilde{\lambda}_{rs}$.

Reliability of the sth component of rth branch is as follows:

$$\tilde{R}_{rs}(t) = \exp\left(- \tilde{\lambda}_{rs}.t \right)$$

It is well known that the fuzzy reliability of system $\bar{R}_{PS}(t)$ is given by

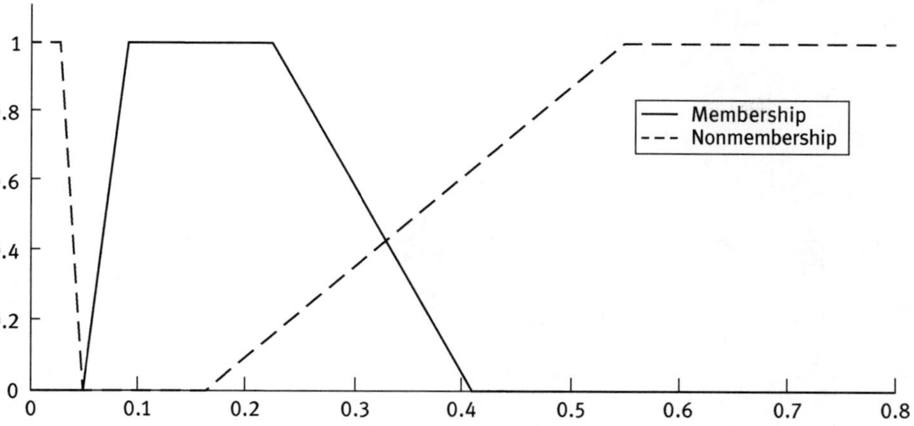

Figure 10.8: For $t = 30$, membership function and nonmembership function of reliability of achieving orbit.

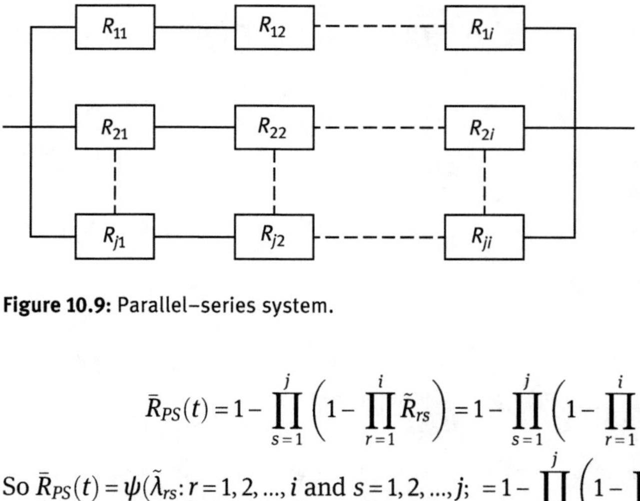

Figure 10.9: Parallel–series system.

$$\bar{R}_{PS}(t) = 1 - \prod_{s=1}^{j}\left(1 - \prod_{r=1}^{i}\tilde{R}_{rs}\right) = 1 - \prod_{s=1}^{j}\left(1 - \prod_{r=1}^{i}\exp(-\tilde{\lambda}_{rs}.t)\right) \tag{10.17}$$

So $\bar{R}_{PS}(t) = \psi(\tilde{\lambda}_{rs}: r = 1, 2, ..., i \text{ and } s = 1, 2, ..., j; = 1 - \prod_{s=1}^{j}\left(1 - \prod_{r=1}^{i}\exp(-\tilde{\lambda}_{rs}.t)\right)$

α-Cut of $\tilde{\lambda}_{rs}$ for membership function is

$$\tilde{\lambda}_{rs\alpha} = [h^L_{rs\alpha}, h^R_{rs\alpha}] \qquad \forall\alpha \in [0, 1], r = 1, 2, ..., i \text{ and } s = 1, 2, ..., j$$

For each α, we have n^2 intervals $\tilde{\lambda}_{11\alpha} = [h^L_{11\alpha}, h^R_{11\alpha}]$, $\tilde{\lambda}_{12\alpha} = [h^L_{12\alpha}, h^R_{12\alpha}]$, $\tilde{\lambda}_{21\alpha} = [h^L_{21\alpha}, h^R_{21\alpha}]$, . . . , $\tilde{\lambda}_{ij\alpha} = [h^L_{ij\alpha}, h^R_{ij\alpha}]$, for each $\tilde{\lambda}_{rs}$, $1 \le r \le i$ and $1 \le s \le j$, respectively.

For nonmembership function, β-Cut of $\tilde{\lambda}_{rs}$ is

$$\tilde{\lambda}_{rs\beta} = [h^L_{rs\beta}, h^R_{rs\beta}] \qquad \forall\beta \in [0, 1], r = 1, 2, ..., i \text{ and } s = 1, 2, ..., j$$

Similarly, for every β, we have n^2 intervals $\tilde{\lambda}_{11\beta} = [h^L_{11\beta}, h^R_{11\beta}]$, $\tilde{\lambda}_{12\beta} = [h^L_{12\beta}, h^R_{12\beta}]$, $\tilde{\lambda}_{21\beta} = [h^L_{21\beta}, h^R_{21\beta}]$, . . . , $\tilde{\lambda}_{ij\beta} = [h^L_{ij\beta}, h^R_{ij\beta}]$, for each $\tilde{\lambda}_{rs}$, $1 \le r \le i$ and $1 \le s \le j$, respectively.

α-Cut of $\tilde{R}_{PS}(t)$ for membership function is

$$\tilde{R}_{PS\alpha}(t) = [\psi^L_{PS\alpha}(t), \psi^R_{PS\alpha}(t)]$$

where

$$\psi^L_{PS\alpha}(t) = \min \left\{1 - \prod_{s=1}^{j}\left(1 - \prod_{r=1}^{i} \exp(-\tilde{\lambda}_{rs}.t)\right)\right\}, \quad \psi^L_{PS\alpha}(t) = \min\left\{1 - \prod_{s=1}^{j}\left(1 - \prod_{r=1}^{i} \exp(-\tilde{\lambda}_{rs}.t)\right)\right\}$$
$$\text{s.t. } h^L_{11\alpha} \le x_{11} \le h^R_{11\alpha}, h^L_{12\alpha} \le x_{12} \le h^R_{12\alpha}, h^L_{21\alpha} \le x_{21} \le h^R_{21\alpha}, \ldots, h^L_{ij\alpha} \le x_{ij} \le h^R_{ij\alpha} \tag{10.18}$$

β-Cut of $\tilde{R}_{PS}(t)$ for nonmembership function is

$$\tilde{R}_{PS\beta}(t) = [\psi^L_{PS\beta}(t), \psi^R_{PS\beta}(t)]$$

where

$$\psi^L_{PS\beta}(t) = \min \left\{1 - \prod_{s=1}^{j}\left(1 - \prod_{r=1}^{i} \exp(-\tilde{\lambda}_{rs}.t)\right)\right\}, \quad \psi^L_{PS\beta}(t) = \min\left\{1 - \prod_{s=1}^{j}\left(1 - \prod_{r=1}^{i} \exp(-\tilde{\lambda}_{rs}.t)\right)\right\}$$
$$\text{s.t. } h^L_{11\beta} \le x_{11} \le h^R_{11\beta}, h^L_{12\beta} \le x_{12} \le h^R_{12\beta}, h^L_{21\beta} \le x_{21} \le h^R_{21\beta}, \ldots, h^L_{ij\beta} \le x_{ij} \le h^R_{ij\beta} \tag{10.19}$$

After solving (10.18) and (10.19), we calculate α and β-Cut of fuzzy reliability of above system $\tilde{R}_{PS}(t)$.

10.6.1 Numerical example

Suppose there is a communication system which receives the input signal and transmits the output signal. For this, there are two receivers and two transmitters in the system connected as is in Figure 10.10. For a successful communication at least one receiver and one transmitter connected in series configuration must work properly.

The failure rates of receivers and transmitters are in the form of different types of CBFNs given in Table 10.9.

The α-Cuts of $\tilde{\lambda}_{11}, \tilde{\lambda}_{12}, \tilde{\lambda}_{21}, \tilde{\lambda}_{22}$ are computed and tabulated in Table 10.10.

The β-Cut of $\tilde{\lambda}_{11}, \tilde{\lambda}_{12}, \tilde{\lambda}_{21}, \tilde{\lambda}_{22}$ are computed and tabulated in Table 10.11.

Using formulation (10.18) and (10.19), we get

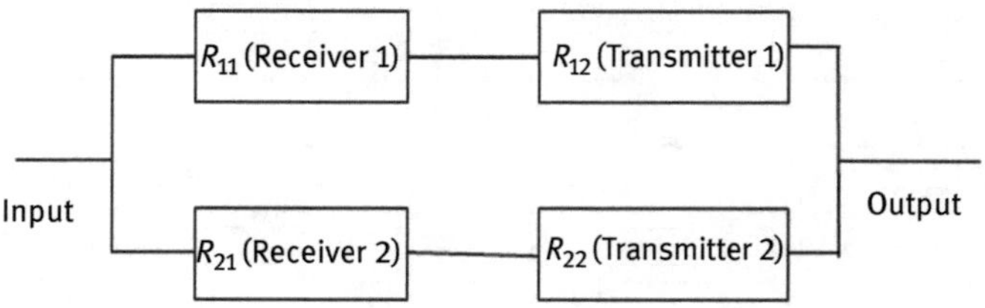

Figure 10.10: Parallel–series system.

Table 10.9: Conflicting bifuzzy failure rates of components.

Bifuzzy failure rate	Types of bifuzzy number
$\tilde{\lambda}_{11} = (0.05,\ 0.10,\ 0.15;\ 0.04,\ 0.12,\ 0.18)$	Triangular conflicting bifuzzy number
$\tilde{\lambda}_{12} = (0.06,\ 0.09,\ 0.12,\ 0.16;\ 0.07,\ 0.11,\ 0.15,\ 0.18)$	Trapezoidal conflicting bifuzzy number
$\tilde{\lambda}_{21} = (0.07,\ 0.12,\ 0.17;\ 0.09,\ 0.14,\ 0.18)$	Triangular conflicting bifuzzy number
$\tilde{\lambda}_{22} = (0.03,\ 0.08,\ 0.13,\ 018;\ 0.05,\ 0.10,\ 0.18,\ 0.23)$	Trapezoidal conflicting bifuzzy number

Table 10.10: α-Cut of $\tilde{\lambda}_{11}$, $\tilde{\lambda}_{12}$, $\tilde{\lambda}_{21}$, and $\tilde{\lambda}_{22}$ for membership.

α	$\tilde{\lambda}_{11\alpha}$	$\tilde{\lambda}_{12\alpha}$	$\tilde{\lambda}_{21\alpha}$	$\tilde{\lambda}_{22\alpha}$
0.0	[0.05, 0.15]	[0.06, 0.16]	[0.07, 0.17]	[0.03, 0.18]
0.1	[0.055, 0.145]	[0.063, 0.156]	[0.075, 0.165]	[0.035, 0.175]
0.2	[0.06, 0.14]	[0.066, 0.152]	[0.08, 0.16]	[0.04, 0.17]
0.3	[0.065, 0.135]	[0.069, 0.148]	[0.085, 0.155]	[0.045, 0.165]
0.4	[0.07, 0.13]	[0.072, 0.144]	[0.09, 0.15]	[0.05, 0.16]
0.5	[0.075, 0.125]	[0.075, 0.140]	[0.095, 0.145]	[0.055, 0.155]
0.6	[0.08, 0.12]	[0.078, 0.136]	[0.10, 0.14]	[0.06, 0.15]
0.7	[0.085, 0.115]	[0.081, 0.132]	[0.105, 0.135]	[0.065, 0.145]
0.8	[0.09, 0.11]	[0.084, 0.128]	[0.11, 0.13]	[0.07, 0.14]
0.9	[0.095, 0.105]	[0.087, 0.124]	[0.115, 0.125]	[0.075, 0.135]
1.0	[0.10, 0.10]	[0.09, 0.12]	[0.12, 0.12]	[0.08, 0.13]

Table 10.11: β-Cut of $\tilde{\lambda}_{11}$, $\tilde{\lambda}_{12}$, $\tilde{\lambda}_{21}$, and $\tilde{\lambda}_{22}$ for nonmembership.

β	$\tilde{\lambda}_{11\beta}$	$\tilde{\lambda}_{12\beta}$	$\tilde{\lambda}_{21\beta}$	$\tilde{\lambda}_{22\alpha}$
0.0	[0.12, 0.12]	[0.11, 0.15]	[0.14, 0.14]	[0.10, 0.18]
0.1	[0.112, 0.126]	[0.106, 0.153]	[0.135, 0.144]	[0.095, 0.185]
0.2	[0.104, 0.132]	[0.102, 0.156]	[0.130, 0.148]	[0.09, 0.19]
0.3	[0.096, 0.138]	[0.098, 0.159]	[0.125, 0.152]	[0.085, 0.195]
0.4	[0.088, 0.144]	[0.094, 0.162]	[0.120, 0.156]	[0.08, 0.20]
0.5	[0.08, 0.15]	[0.090, 0.165]	[0.115, 0.160]	[0.075, 0.205]
0.6	[0.072, 0.156]	[0.086, 0.168]	[0.110, 0.164]	[0.07, 0.21]
0.7	[0.064, 0.162]	[0.082, 0.171]	[0.105, 0.168]	[0.065, 0.215]
0.8	[0.056, 0.168]	[0.078, 0.174]	[0.10, 0.172]	[0.06, 0.22]
0.9	[0.048, 0.174]	[0.074, 0.177]	[0.095, 0.176]	[0.055, 0.225]
1.0	[0.04, 0.18]	[0.07, 0.18]	[0.09, 0.18]	[0.05, 0.23]

$$
\begin{aligned}
\psi^{L}_{PS\alpha}(t) &= 1 - \{(1 - e^{-(.36-.1\alpha)t}) \times (1 - e^{-(.34-.09\alpha)t})\} \\
\psi^{R}_{PS\alpha}(t) &= 1 - \{(1 - e^{-(.12+.1\alpha)t}) \times (1 - e^{-(.09+.08\alpha)t})\} \\
\psi^{L}_{PS\beta}(t) &= 1 - \{(1 - e^{-(.30+.1\beta)t}) \times (1 - e^{-(.33+.08\beta)t})\} \\
\psi^{R}_{PS\beta}(t) &= 1 - \{(1 - e^{-(.26-.13\beta)t}) \times (1 - e^{-(.21-.09\beta)t})\}
\end{aligned}
\tag{10.20}
$$

The fuzzy system reliability of given system is evaluated using (10.20). The fuzzy reliability of system is trapezoidal CBFN for failure rate of components follow conflicting bifuzzy. The α-Cut and β-Cut of fuzzy reliability $\bar{R}_{PS}(t)$ are evaluated for $t =$ 10, 20, 30 and shown in Table 10.12. The value of sum of membership and nonmembership degree is found to be greater than one (Figures 10.11–10.13) at point A, B, C.

10.7 Reliability of series—parallel system when failure rates follow conflicting bifuzzy numbers

Let us assume a series—parallel system consisting "" subsystems connected in series and each subsystem contains "j" components connected in parallel form as shown in Figure 10.14. Let failure rates of all components are assumed to be different CBFNs $\tilde{\lambda}_{rs}$.

Table 10.12: α-Cut and β-Cut of parallel–series system reliability $\bar{R}_{PS}(t)$ for $t = 10, 20, 30$.

α, β	For $t = 10$	For $t = 20$	For $t = 30$
	α-Cut β-Cut	α-Cut β-Cut	α-Cut β-Cut
0	[.0597, .5853] [.0848, .1876]	[.0018, .2410] [.0038, .0204]	[.00005, .09269] [.00017, .00225]
1	[.0656, .5456] [.0776, .2072]	[.0022, .2046] [.0032, .0249]	[.00008, .07204] [.00013, .00301]
2	[.0719, .5076] [.0709, .2288]	[.0027, .1735] [.0027, .0306]	[.00010, .05596] [.00009, .00404]
3	[.0789, .4716] [.0648, .2525]	[.0033, .1469] [.0022, .0375]	[.00013, .04346] [.00007, .00544]
4	[.0866, .4375] [.0593, .2785]	[.0039, .1244] [.0018, .0459]	[.00018, .03375] [.00005, .00735]
5	[.0950, .4054] [.0542, .3070]	[.0047, .1052] [.0015, .0564]	[.00024, .02622] [.00004, .00994]
6	[.1042, .3753] [.0495, .3381]	[.0057, .0888] [.0013, .0693]	[.00031, .02037] [.00003, .01349]
7	[.1142, .3471] [.0453, .3720]	[.0069, .0751] [.0011, .0851]	[.00041, .01583] [.00003, .01836]
8	[.1252, .3207] [.0414, .4088]	[.0084, .0634] [.0009, .1047]	[.00055, .01231] [.00002, .02505]
9	[.1372, .2962] [.0378, .4487]	[.0101, .0536] [.0007, .1287]	[.00073, .00957] [.00001, .03428]
1	[.1503, .2734] [.0346, .4916]	[.0122, .0452] [.0006, .1583]	[.00096, .00745] [.00001, .04701]

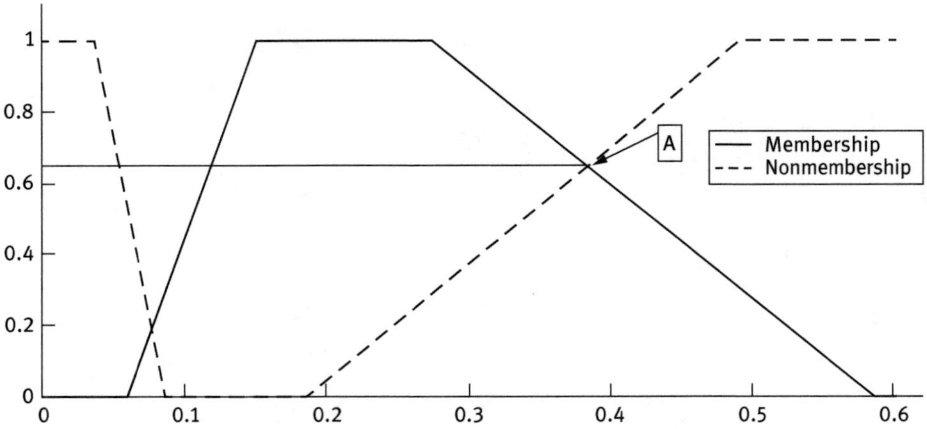

Figure 10.11: For $t = 10$, membership function and nonmembership function of reliability of communication system.

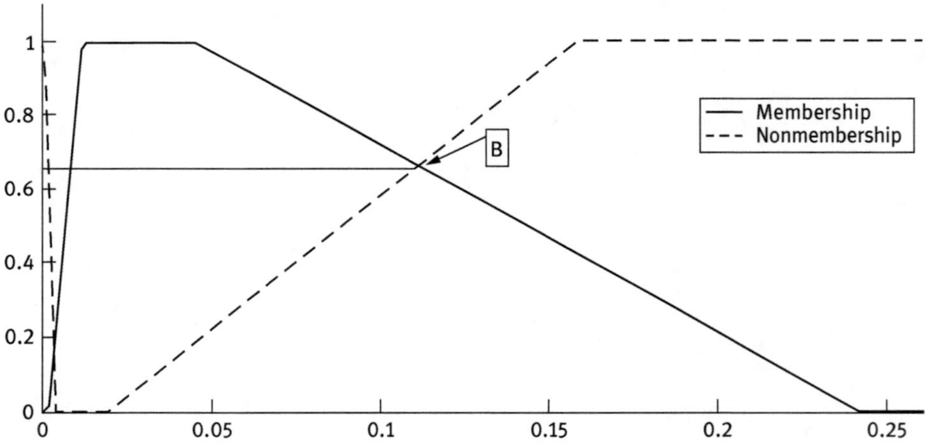

Figure 10.12: For $t = 20$, membership function and nonmembership function of reliability of communication system.

Let failure rate function of the rth component of sth subsystem ($r = 1, 2, \ldots, j$ and $s = 1, 2, \ldots, i$) is represented by CBFS $\bar{\lambda}_{rs}$.

Reliability of the rth component of sth branch is as follows:

$$\bar{R}_{rs}(t) = \exp\left(-\bar{\lambda}_{rs}.t\right)$$

It is well known that the fuzzy reliability of system $\bar{R}_{SP}(t)$ is given below:

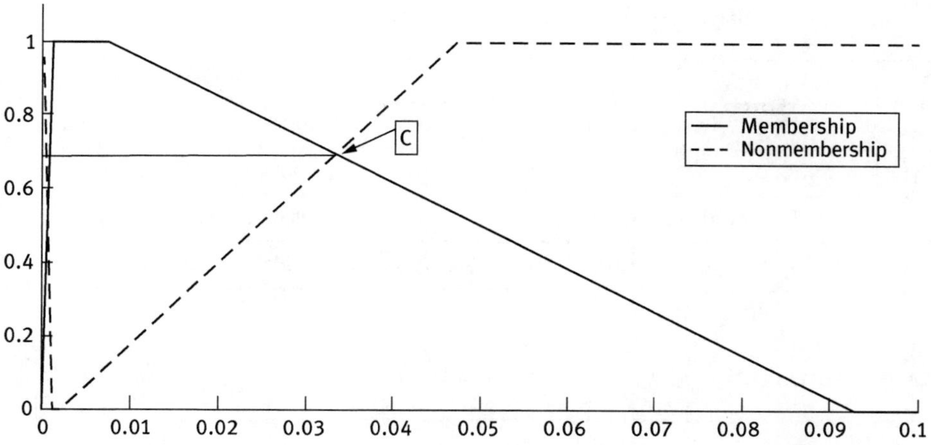

Figure 10.13: For $t = 30$, membership function and nonmembership function of reliability of communication system.

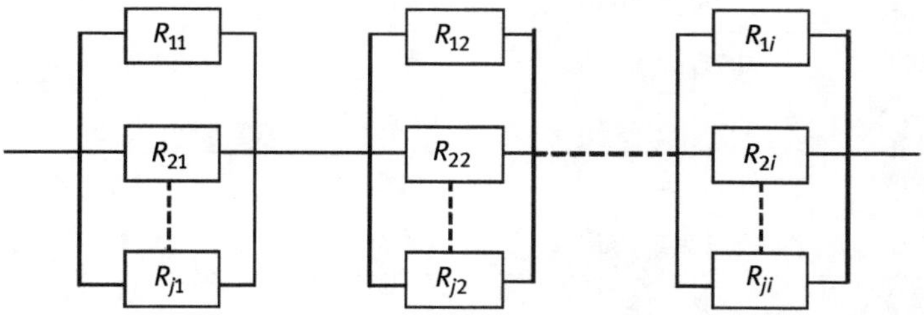

Figure 10.14: Series–parallel system.

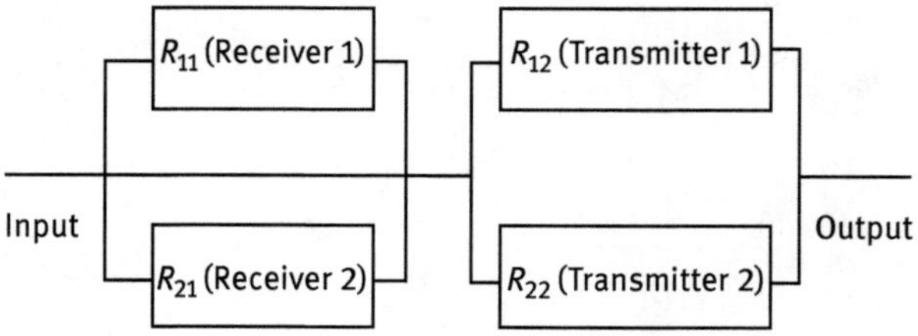

Figure 10.15: Series–parallel system.

Table 10.13: Conflicting bifuzzy failure rates of components.

Bifuzzy failure rate	Types of bifuzzy number
$\tilde{\lambda}_{11} = (0.06\ ,0.09,\ 0.12;\ 0.07,\ 0.11,\ 0.16)$	Triangular conflicting bifuzzy number
$\tilde{\lambda}_{12} = (0.07,\ 0.12,\ 0.17,\ 0.22;\ 0.09,\ 0.14,\ 0.19,\ 0.24)$	Trapezoidal conflicting bifuzzy number
$\tilde{\lambda}_{21} = (0.06,\ 0.11,\ 0.16;\ 0.08,\ 0.13,\ 0.18)$	Triangular conflicting bifuzzy number
$\tilde{\lambda}_{22} = (0.08,\ 0.13,\ 0.18,\ 0.22;\ 0.10,\ 0.15,\ 0.20,\ 0.25)$	Trapezoidal conflicting bifuzzy number

Table 10.14: α-Cut of $\tilde{\lambda}_{11}$, $\tilde{\lambda}_{12}$, $\tilde{\lambda}_{21}$, and $\tilde{\lambda}_{22}$ for membership.

α	$\tilde{\lambda}_{11\alpha}$	$\tilde{\lambda}_{12\alpha}$	$\tilde{\lambda}_{21\alpha}$	$\tilde{\lambda}_{22\alpha}$
0.0	[0.06, 0.12]	[0.07, 0.22]	[0.06, 0.16]	[0.08, 0.22]
0.1	[0.063, 0.117]	[0.075, 0.215]	[0.065, 0.155]	[0.085, 0.216]
0.2	[0.066, 0.114]	[0.08, 0.21]	[0.07, 0.15]	[0.090, 0.212]
0.3	[0.069, 0.111]	[0.085, 0.205]	[0.075, 0.145]	[0.095, 0.208]
0.4	[0.072, 0.108]	[0.09, 0.20]	[0.08, 0.14]	[0.10, 0.204]
0.5	[0.075, 0.105]	[0.095, 0.195]	[0.085, 0.135]	[0.105, 0.20]
0.6	[0.078, 0.102]	[0.10, 0.19]	[0.09, 0.13]	[0.11, 0.196]
0.7	[0.081, 0.099]	[0.105, 0.185]	[0.095, 0.125]	[0.115, 0.192]
0.8	[0.084, 0.096]	[0.11, 0.18]	[0.10, 0.12]	[0.12, 0.188]
0.9	[0.087, 0.093]	[0.115, 0.175]	[0.105, 0.115]	[0.125, 0.184]
1.0	[0.09, 0.09]	[0.12, 0.17]	[0.11, 0.11]	[0.13, 0.18]

$$\bar{R}_{SP} = \prod_{s=1}^{i}\left(1 - \prod_{r=1}^{j}\left(1 - \bar{R}_{rs}\right)\right) = \prod_{s=1}^{i}\left(1 - \prod_{r=1}^{j}\left(1 - \exp(-\tilde{\lambda}_{rs})\right)\right) \tag{10.21}$$

So $\bar{R}_{SP}(t) = \psi(\tilde{\lambda}_{rs}: r = 1, 2, ..., j$ and $s = 1, 2, ..., i;\ t = \prod_{s=1}^{i}\{1 - \prod_{r=1}^{j}(1 - \exp(-\tilde{\lambda}_{rs}))\}$

α-Cut of $\tilde{\lambda}_{rs}$ for membership function is

$$\tilde{\lambda}_{rs\alpha} = [h^L_{rs\alpha}, h^R_{rs\alpha}] \quad \forall \alpha \in [0, 1], r = 1, 2, ..., j \text{ and } s = 1, 2, ..., i.$$

For every α, we have n^2 intervals $\tilde{\lambda}_{11\alpha} = [h^L_{11\alpha}, h^R_{11\alpha}]$, $\tilde{\lambda}_{12\alpha} = [h^L_{12\alpha}, h^R_{12\alpha}]$, $\tilde{\lambda}_{21\alpha} = [h^L_{21\alpha}, h^R_{21\alpha}]$, ..., $\tilde{\lambda}_{ji\alpha} = [h^L_{ji\alpha}, h^R_{ji\alpha}]$, for each $\tilde{\lambda}_{rs}$, $1 \le r \le j$ and $1 \le s \le i$, respectively.

For nonmembership function, β-Cut of $\tilde{\lambda}_{rs}$ is

Table 10.15: β-Cut of $\tilde{\lambda}_{11}$,$\tilde{\lambda}_{12}$,$\tilde{\lambda}_{21}$, and $\tilde{\lambda}_{22}$ for nonmembership.

β	$\tilde{\lambda}_{11\beta}$	$\tilde{\lambda}_{12\beta}$	$\tilde{\lambda}_{21\beta}$	$\tilde{\lambda}_{22\beta}$
0.0	[0.11, 0.11]	[0.14, 0.19]	[0.13, 0.13]	[0.15, 0.20]
0.1	[0.106, 0.115]	[0.135, 0.195]	[0.125, 0.135]	[0.145, 0.205]
0.2	[0.102, 0.120]	[0.13, 0.20]	[0.12, 0.14]	[0.14, 0.21]
0.3	[0.098, 0.125]	[0.125, 0.205]	[0.115, 0.145]	[0.135, 0.215]
0.4	[0.094, 0.130]	[0.12, 0.21]	[0.11, 0.15]	[0.13, 0.22]
0.5	[0.090, 0.135]	[0.115, 0.215]	[0.105, 0.155]	[0.125, 0.225]
0.6	[0.086, 0.140]	[0.11, 0.22]	[0.10, 0.16]	[0.12, 0.23]
0.7	[0.082, 0.145]	[0.105, 0.225]	[0.095, 0.165]	[0.115, 0.235]
0.8	[0.078, 0.150]	[0.10, 0.23]	[0.09, 0.17]	[0.11, 0.24]
0.9	[0.074, 0.155]	[0.095, 0.235]	[0.085, 0.175]	[0.105, 0.245]
1.0	[0.07, 0.16]	[0.09, 0.24]	[0.08, 0.18]	[0.10, 0.25]

$$\tilde{\lambda}_{rs\beta} = [h^L_{rs\beta}, h^R_{rs\beta}] \quad \forall \beta \in [0,1], r = 1, 2, ..., j \text{ and } s = 1, 2, ..., i$$

Similarly, for each β, we have n^2 intervals $\tilde{\lambda}_{11\beta} = [h^L_{11\beta}, h^R_{11\beta}]$, $\tilde{\lambda}_{12\beta} = [h^L_{12\beta}, h^R_{12\beta}]$, $\tilde{\lambda}_{21\beta} = [h^L_{21\beta}, h^R_{21\beta}]$, . . . , $\tilde{\lambda}_{ji\beta} = [h^L_{ji\beta}, h^R_{ji\beta}]$, for any $\tilde{\lambda}_{rs}$, $1 \le r \le j$ and $1 \le s \le i$, respectively.

α-Cut of $\tilde{R}_{SP}(t)$ for membership function is

$$\tilde{R}_{SP\alpha}(t) = [\psi^L_{SP\alpha}(t), \psi^R_{SP\alpha}(t)]$$

where

$$\psi^L_{SP\alpha}(t) = \min \left\{ \prod_{s=1}^{i} \left(1 - \prod_{r=1}^{j} \left(1 - \exp(-\tilde{\lambda}_{rs})\right)\right)\right\}, \ \psi^L_{SP\alpha}(t) = \min\left\{ \prod_{s=1}^{i} \left(1 - \prod_{r=1}^{j} \left(1 - \exp(-\tilde{\lambda}_{rs})\right)\right)\right\}$$
$$\text{s.t. } h^L_{11\alpha} \le x_{11} \le h^R_{11\alpha}, h^L_{12\alpha} \le x_{12} \le h^R_{12\alpha}, h^L_{21\alpha} \le x_{21} \le h^R_{21\alpha}, ..., h^L_{ij\alpha} \le x_{ij} \le h^R_{ij\alpha} \tag{10.22}$$

β-Cut of $\tilde{R}_{SP}(t)$ for nonmembership function is

$$\tilde{R}_{SP\beta}(t) = [\psi^L_{SP\beta}(t), \psi^R_{SP\beta}(t)]$$

where

$$\psi^L_{SP\beta}(t) = \min$$

Table 10.16: α and β-Cut of series–parallel system reliability $\tilde{R}_{sp}(t)$ for $t = 10, 20, 30$.

α, β	For $t = 10$	For $t = 20$	For $t = 30$
	α-Cut β-Cut	α-Cut β-Cut	α-Cut β-Cut
0	[.0926, .5755] [.1362, .2134]	[.0031, .2040] [.00712, .01902]	[.00009, .06128] [.00033, .00146]
1	[.9974, .5419] [.1248, .2311]	[.0037, .1748] [.00586, .02279]	[.00012, .04764] [.00024, .00193]
2	[.1074, .5095] [.1142, .2499]	[.0043, .1495] [.00482, .02728]	[.00016, .03698] [.00018, .00255]
3	[.1157, .4782] [.1045, .2702]	[.0050, .1276] [.00397, .03264]	[.00019, .02868] [.00013, .00336]
4	[.1245, .4483] [.0956, .2918]	[.0059, .1088] [.00326, .03901]	[.00025, .02223] [.00009, .00443]
5	[.1339, .4197] [.0873, .3148]	[.0069, .0926] [.00268, .04658]	[.00032, .01721] [.00007, .00584]
6	[.1441, .3924] [.0797, .3392]	[.0081, .0788] [.00220, .05556]	[.00041, .01332] [.00006, .00769]
7	[.1549, .3666] [.0728, .3651]	[.0095, .0668] [.00181, .06619]	[.00052, .01031] [.00004, .01013]
8	[.1665, .3421] [.0664, .3924]	[.0112, .0568] [.00148, .07875]	[.00066, .00797] [.00003, .01332]
9	[.1788, .3188] [.0606, .4213]	[.0131, .0481] [.00122, .09355]	[.00084, .00617] [.00002, .01750]
1	[.1919, .2971] [.0552, .4516]	[.0154, .0408] [.00099, .11095]	[.00108, .00478] [.00002, .02296]

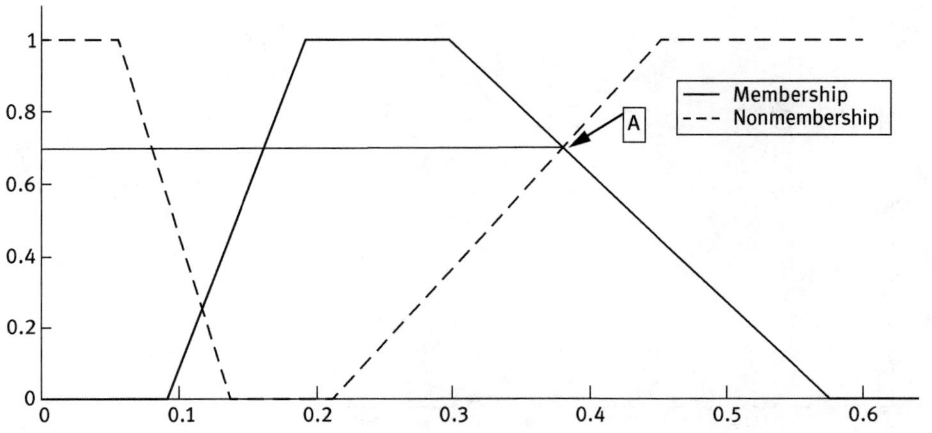

Figure 10.16: For $t = 10$, membership function and nonmembership function of reliability of communication system.

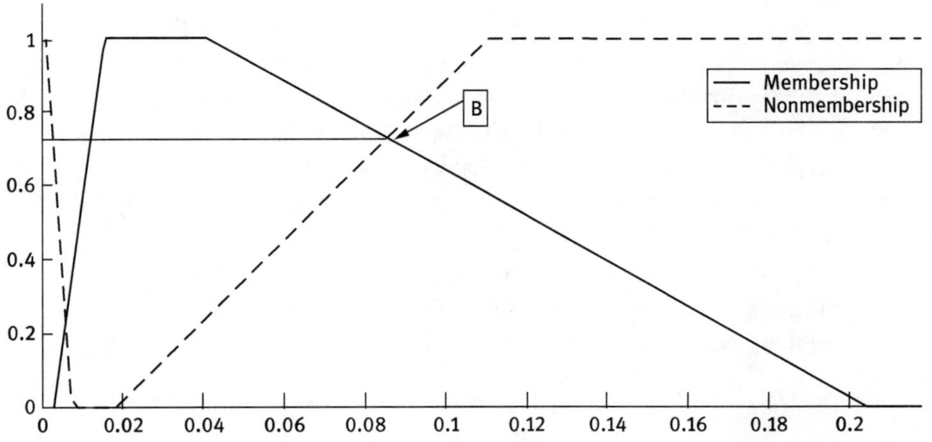

Figure 10.17: For $t = 20$, membership function and nonmembership function of reliability of communication system.

$$\left\{\prod_{s=1}^{i}\left(1-\prod_{r=1}^{j}\left(1-\exp(-\tilde{\lambda}_{rs})\right)\right)\right\},\ \psi_{SP\beta}^{L}(t)=\min\left\{\prod_{s=1}^{i}\left(1-\prod_{r=1}^{j}\left(1-\exp(-\tilde{\lambda}_{rs})\right)\right)\right\}$$
$$\text{s.t. } h_{11\beta}^{L}\le x_{11}\le h_{11\beta}^{R}, h_{11\beta}^{L}\le x_{11}\le h_{11\beta}^{R}, h_{11\beta}^{L}\le x_{11}\le h_{11\beta}^{R}, \ldots, h_{11\beta}^{L}\le x_{11}\le h_{11\beta}^{R} \qquad (10.23)$$

After solving (10.21) and (10.22), we get out α-Cut and β-Cut of fuzzy reliability of series—parallel system $\tilde{R}_{SP}(t)$.

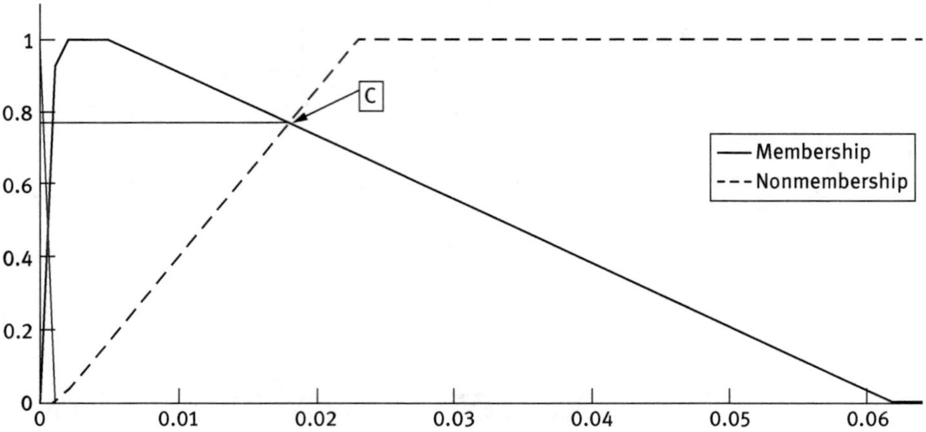

Figure 10.18: For $t = 30$, membership function and nonmembership function of reliability of communication system.

10.7.1 Numerical problem

Suppose there is a communication system which receives the input signal and transmits the output signal. For this there are two receivers and two transmitters in the system connected as shown in Figure 10.15. For a successful communication at least one receiver and one transmitter must work properly.

The failure rate of receivers and transmitters follow different types of CBFNs as shown in Table 10.13.

The α-Cut of $\tilde{\lambda}_{11}$, $\tilde{\lambda}_{12}$, $\tilde{\lambda}_{21}$, $\tilde{\lambda}_{22}$ are calculated and tabulated in Table 10.14.

The β-Cuts of $\tilde{\lambda}_{11}$, $\tilde{\lambda}_{12}$, $\tilde{\lambda}_{21}$, $\tilde{\lambda}_{22}$ are calculated and tabulated in Table 10.15.

Using formulae (10.21) and (10.22), we get

$$\begin{aligned}
\psi^L_{SP\alpha}(t) &= \{1-(1-e^{-(0.12-0.03\alpha)t}) \times (1-e^{-(0.16-0.05\alpha)t})\} \times \{1-(1-e^{-(0.22-0.05\alpha)t}) \\
&\quad \times (1-e^{-(0.22-0.04\alpha)t})\} \\
\psi^R_{SP\alpha}(t) &= \{1-(1-e^{-(0.06+0.03\alpha)t}) \times (1-e^{-(0.06+0.05\alpha)t})\} \times \{1-(1-e^{-(0.07+0.05\alpha)t}) \\
&\quad \times (1-e^{-(0.08+0.05\alpha)t})\} \\
\psi^L_{SP\beta}(t) &= \{1-(1-e^{-(0.11+0.05\beta)t}) \times (1-e^{-(0.13+0.05\beta)t})\} \times \{1-(1-e^{-(0.19+0.05\beta)t}) \\
&\quad \times (1-e^{-(0.20+0.05\beta)t})\} \\
\psi^R_{SP\beta}(t) &= \{1-(1-e^{-(0.11-0.04\beta)t}) \times (1-e^{-(0.13-0.05\beta)t})\} \times \{1-(1-e^{-(0.14-0.05\beta)t}) \\
&\quad \times (1-e^{-(0.15-0.05\beta)t})\}
\end{aligned} \tag{10.24}$$

The system reliability of above system is calculated using (10.24). The system reliability is in trapezoidal CBFN for failure rate of components follow conflicting bifuzzy. The α-Cut and β-Cut of fuzzy reliability of above system $\bar{R}_{SP}(t)$ are evaluated for $t = 10, 20, 30$ and given in Table 10.16. The sum of membership and nonmembership degree is found to be greater than one (Figures 10.16–10.18) at points A, B, C.

10.8 Conclusion

In this study, we have discussed a new method for evaluating fuzzy reliability of different systems having components follow different types of conflicting bifuzzy failure rates. We also fabricate the membership degree and nonmembership degree of fuzzy reliability Kumar and Yadav [9]. Further, using introduced method, membership degree and nonmembership degree of fuzzy reliability of different systems like series system, parallel system, parallel–series system, and series–parallel system. The components of above systems follow different types of conflicting bifuzzy failure rates. We have seen that, the fuzzy reliabilities of above systems are in the form of trapezoidal conflicting bifuzzy numbers. The α-Cut and β-Cut of fuzzy reliability are also evaluated at different times (t).

References

[1] Zadeh L.A. (1965) Fuzzy sets, Informatics and Control, 8(3), 338–353.

[2] Zamali T., Mohd Lazim A., & Abu Osman M.T. (2008) An introduction to conflicting bifuzzy sets theory, International Journal of Mathematics and Statistics, 86–101.

[3] Imran C.T., Syibrah M.N., & Mohd Lazim A. (2008) A new condition for conflicting bifuzzy sets based on intuitionistic evaluation, World Academy of Science, Engineering and Technology I: Mathematical and Computational Sciences, 2(4), 161–165.

[4] Cai K.Y., Wen C.Y., & Zhang M.L. (1991) Fuzzy variables as a basic for a theory of fuzzy reliability in the possibility context, Fuzzy Set and Systems, 42, 145–172.

[5] Cai K.Y., Wen C.Y., & Zhang M.L. (1993) Fuzzy states as for a theory of fuzzy reliability, Microelectron Reliability, 33, 2253–2263.

[6] Chen C.H. & Mon L. (1993) Fuzzy system reliability analysis by interval of confidence, Fuzzy Set and Systems, 56, 29–35.

[7] Onisawa T. & Kacprzyk J. (1995) Reliability and safety analysis under fuzziness, Physica-Verlag Heidelberg.

[8] Cai K.Y. (1996) System failure engineering and fuzzy methodology. An introductory overview, Fuzzy Set and Systems, 83, 113–133.

[9] Kumar M. & Yadav S.P. (2012) A novel approach for analyzing fuzzy system reliability using different types of intuitionistic fuzzy failure rates of components, ISA Transactions, 51, 288–297.

[10] Aliev M. & Kara Z. (2004) Fuzzy system reliability analysis using time dependent fuzzy set, Control and Cybernetics, 33, 653–662.

[11] Mon D.L. & Cheng C.H. (1994) Fuzzy system reliability analysis for components with different membership functions, Fuzzy Sets and Systems, 64, 145–157.
[12] Kumar M., Yadav S.P., & Kumar S. (2011) A new approach for analyzing the fuzzy system reliability using intuitionistic fuzzy number, International Journal of Industrial and Systems Engineering, 8(2), 135–156.

Index

https://doi.org/10.1515/9783110628616-011